Recent Titles in This Series

126 **M. L. Agranovskiĭ,** Invariant function spaces on homogeneous manifolds of Lie groups and applications, 1993
125 **Masayoshi Nagata,** Theory of commutative fields, 1993
124 **Masahisa Adachi,** Embeddings and immersions, 1993
123 **M. A. Akivis and B. A. Rosenfeld,** Élie Cartan (1869–1951), 1993
122 **Zhang Guan-Hou,** Theory of entire and meromorphic functions: Deficient and asymptotic values and singular directions, 1993
121 **I. B. Fesenko and S. V. Vostokov,** Local fields and their extensions: A constructive approach, 1993
120 **Takeyuki Hida and Masuyuki Hitsuda,** Gaussian processes, 1993
119 **M. V. Karasev and V. P. Maslov,** Nonlinear Poisson brackets. Geometry and quantization, 1993
118 **Kenkichi Iwasawa,** Algebraic functions, 1993
117 **Boris Zilber,** Uncountably categorical theories, 1993
116 **G. M. Fel′dman,** Arithmetic of probability distributions, and characterization problems on abelian groups, 1993
115 **Nikolai V. Ivanov,** Subgroups of Teichmüller modular groups, 1992
114 **Seizô Itô,** Diffusion equations, 1992
113 **Michail Zhitomirskiĭ,** Typical singularities of differential 1-forms and Pfaffian equations, 1992
112 **S. A. Lomov,** Introduction to the general theory of singular perturbations, 1992
111 **Simon Gindikin,** Tube domains and the Cauchy problem, 1992
110 **B. V. Shabat,** Introduction to complex analysis Part II. Functions of several variables, 1992
109 **Isao Miyadera,** Nonlinear semigroups, 1992
108 **Takeo Yokonuma,** Tensor spaces and exterior algebra, 1992
107 **B. M. Makarov, M. G. Goluzina, A. A. Lodkin, and A. N. Podkorytov,** Selected problems in real analysis, 1992
106 **G.-C. Wen,** Conformal mappings and boundary value problems, 1992
105 **D. R. Yafaev,** Mathematical scattering theory: General theory, 1992
104 **R. L. Dobrushin, R. Kotecký, and S. Shlosman,** Wulff construction: A global shape from local interaction, 1992
103 **A. K. Tsikh,** Multidimensional residues and their applications, 1992
102 **A. M. Il′in,** Matching of asymptotic expansions of solutions of boundary value problems, 1992
101 **Zhang Zhi-fen, Ding Tong-ren, Huang Wen-zao, and Dong Zhen-xi,** Qualitative theory of differential equations, 1992
100 **V. L. Popov,** Groups, generators, syzygies, and orbits in invariant theory, 1992
99 **Norio Shimakura,** Partial differential operators of elliptic type, 1992
98 **V. A. Vassiliev,** Complements of discriminants of smooth maps: Topology and applications, 1992
97 **Itiro Tamura,** Topology of foliations: An introduction, 1992
96 **A. I. Markushevich,** Introduction to the classical theory of Abelian functions, 1992
95 **Guangchang Dong,** Nonlinear partial differential equations of second order, 1991
94 **Yu. S. Il′yashenko,** Finiteness theorems for limit cycles, 1991
93 **A. T. Fomenko and A. A. Tuzhilin,** Elements of the geometry and topology of minimal surfaces in three-dimensional space, 1991
92 **E. M. Nikishin and V. N. Sorokin,** Rational approximations and orthogonality, 1991

(Continued in the back of this publication)

Invariant Function Spaces on Homogeneous Manifolds of Lie Groups and Applications

Translations of

MATHEMATICAL MONOGRAPHS

Volume 126

Invariant Function Spaces on Homogeneous Manifolds of Lie Groups and Applications

M. L. Agranovskiĭ

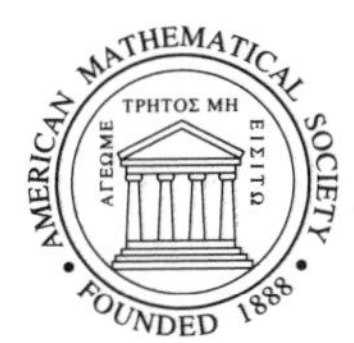

American Mathematical Society
Providence, Rhode Island

ИНВАРИАНТНЫЕ ПРОСТРАНСТВА ФУНКЦИЙ НА ОДНОРОДНЫХ МНОГООБРАЗИЯХ ГРУПП ЛИ И ИХ ПРИЛОЖЕНИЯ

Марк Львович Аграновский

Translated by A. I. Zaslavsky from an original Russian manuscript
Translation edited by D. Louvish

The translation, editing, and keyboarding of the material for this book was done in the framework of the joint project between the AMS and Tel-Aviv University, Israel.

1991 *Mathematics Subject Classification*. Primary 43A85;
Secondary 22E46.

ABSTRACT. The monograph is devoted to the study of translation-invariant function spaces and algebras on homogeneous manifolds: semisimple and nilpotent Lie groups, Riemann symmetric spaces, bounded symmetric domains. Classification of various classes of translation-invariant spaces and algebras is obtained. Applications to characterization problems for holomorphic functions in one and several complex variables and their boundary values are given.

The monograph is directed to specialists in harmonic analysis, function theory, functional analysis, and representation theory of Lie groups.

Library of Congress Cataloging-in-Publication Data

Agranovskiĭ, M. L. (Mark L′vovich)
[Invariantnye prostranstva funk͡tsiĭ na odnorodnykh mnogoobraziĭakh grupp li i ikh prilozheniĭa. English]
Invariant function spaces on homogeneous manifolds of Lie groups and applications/ M. L. Agranovskiĭ.
p. cm.–(Translations of mathematical monographs; v. 126)
Includes bibliographical references.
ISBN 0-8218-4604-3
1. Function spaces. 2. Harmonic analysis. 3. Semisimple Lie groups. I. Title. II. Series
QA323.A3713 1994 93-2029
515′.73–dc20 CIP

The paper used in this book is acid-free and falls within the guidelines
established to ensure permanence and durability. ♾

Information on Copying and Reprinting can be found at the back of this volume.
This publication was typeset using $\mathcal{A}_{\mathcal{M}}\mathcal{S}$-TeX,
the American Mathematical Society's TeX macro system.

10 9 8 7 6 5 4 3 2 1 97 96 95 94 93

To the memory
of my father

Contents

Introduction

This book is devoted to translation-invariant function spaces on homogeneous manifolds — Lie groups, symmetric spaces, complex symmetric domains.

From a general point of view, spaces invariant under translation operators are main objects of harmonic analysis. At the same time, the harmonic analysis of invariant spaces that arise in analytical theories, combined with the methods of those theories themselves, involves some profound and interesting analytical problems.

A constructive study of invariant spaces in specific function classes often presupposes an analytical description. The specific homogeneous structure of the underlying homogeneous space will then inevitably influence the properties of function spaces that are invariant under the given group action.

The following problems are of central importance:

(1) What is the connection between the algebraic-geometric properties of homogeneous spaces, on the one hand, and the properties of the related invariant function spaces, on the other?

(2) Provide an analytical description and classification of invariant subspaces for a given function class on a homogeneous space.

(3) Characterize the classical spaces of function theory from the point of view of their group symmetries: what are the admissible transformation groups for these spaces?

In the 1950s and 1960s, vigorous developments in the theory of uniform algebras established properties of these algebras connected with group invariance.

Shilov [50], [51] and later de Leeuw and Mirkil [80], [83] revealed the specific role of the $C^{(k)}$ scale in the class of translation-invariant function spaces defined on tori and Euclidean spaces. Subsequently, de Leeuw and Mirkil listed all the rotation-invariant uniformly closed function algebras on the real sphere. Wolf [104] then extended this result to the broad class of compact symmetric spaces, revealing the connection between the symmetry of invariant spaces under complex conjugation on the one hand, and the additional symmetry of the Lie algebra of the isometry group of the underlying symmetric space on the other. Translation-invariant uniform algebras on locally compact abelian groups were studied in detail by de Leeuw and Mirkil [81]. Then Wolf [105], Rider [88], and Gangolli [67] discovered new, nontrivial connections between the algebraic properties of compact Lie groups and invariant algebras on such groups, thus showing how one can describe the algebraic properties (semisimplicity, commutativity) in the language of functional analysis.

Thus, one has a kind of duality between a homogeneous space (Lie group, symmetric space) and corresponding class of invariant function spaces (in particular, algebras): each of these two objects conveys information about the other.

Both the results and the methods of the works just surveyed relate mainly to real analysis. In particular, among the analyzed homogeneous spaces were Euclidean spaces, real spheres, compact symmetric spaces, or Lie groups, and whenever a holomorphic structure appeared, it was an isolated episode. But the consideration of invariant function spaces and function algebras on homogeneous spaces with complex structure leads us into the complex domain; moreover, as follows from the results presented here, a special part is played here by holomorphic function spaces and their boundary values. Let us discuss this in more detail.

An important class of homogeneous complex spaces is that of hermitian symmetric spaces whose group of motions is the group of biholomorphic automorphisms. According to Cartan's theory, hermitian symmetric spaces of noncompact type are analytically equivalent to bounded symmetric domains in $\mathbb{C}^n$. This representation leads to two natural homogeneous spaces of the automorphism group—the domain itself and its Shilov boundary. One result of this dissertation is a classification of biholomorphically invariant spaces and function algebras on them. Our results show that the space of holomorphic functions is to a large extent characterized by its biholomorphic invariance. Hence the classification of biholomorphically invariant spaces and algebras is closely related to the characterization of holomorphic functions and their boundary values.

A few words about earlier results in this field. The subject originated in attempts to find multidimensional analogues of Wermer's theorem on the maximality property of the algebra of boundary values of functions holomorphic in a disk. Gorin [41] proved this property for the algebra $A(T^n)$ of boundary values on the torus of functions holomorphic in a polydisk, in the class of closed subalgebras of the algebra $C(T^n)$ invariant under the group automorphisms of the torus T^n admissible for $A(T^n)$. Gorin conjectured that a similar maximality property is valid for the algebra $A(\partial B^n)$ of boundary values of functions holomorphic in the unit ball B^n, with the group automorphisms of the torus replaced by biholomorphic automorphisms of the ball. The conjecture was proved in [31]. A definitive description of all closed biholomorphically invariant (Möbius-type) algebras on a complex sphere was obtained in [2] (see also [1], [9]) (this result was proved again by Nagel and Rudin [84]).

Our subject goes beyond the study of invariant subspaces in symmetric domains; it covers a variety of interrelated questions connected with the classification of invariant function spaces and function algebras on the classical homogeneous spaces — noncompact Lie groups, symmetric spaces, including hermitian spaces, and homogeneous differentiable manifolds.

Let us briefly review the contents. Chapter I is devoted to invariant spaces and algebras on symmetric spaces of noncompact type. We investigate the relationship between the geometry of a symmetric space and the specific properties of the class of isometry invariant L^2-spaces on it. A separate section is devoted to spaces of maximal ideals of antisymmetric algebras of smooth functions on differentiable manifolds (such algebras are typical in our context). The compact analogue of the results of Chapter I concerning invariant spaces are those of Wolf [104], while those concerning algebras on manifolds exploit ideas similar to those Wermer [102]; [103], Hörmander and Wermer [76], Chirka [49], Freeman [65] which deal with approximations by holomorphic functions on real manifolds, our results have a direct bearing on Gelfand's program for investigation of antisymmetric algebras [36].

In Chapter II we study translation invariant function algebras and function spaces on noncompact Lie groups. In the first section we consider the case of semisimple real Lie groups, i.e., isometry groups of symmetric spaces of noncompact type. We prove that in this case the class of uniform algebras invariant under right and left translations is in fact trivial and may be described in terms of central subgroups (just as in the compact case the number of invariant algebras on the group depends directly on the "degree of commutativity" of the group, cf. [105]). Though the results are parallel to their compact counterpats, the Lie groups in question are not compact, so that there are essential differences both in the statement of the problems and in the solution methods. In view of the familiar contrast between compact and noncompact cases in representation theory, we are thus forced to use quite different ideas and means.

A central topic in Chapter II is the investigation of invariant function spaces and function algebras on the (nilpotent) Heisenberg group (§2). Analysis of the Heisenberg group and related groups has lately been receiving much attention (see Folland and Stein's book about Hardy spaces on homogeneous groups [66] for an extensive bibliography). One reason for the interest in this group from the function-theoretic point of view is its equivalence to the punctured complex sphere, so that the analysis on the Heisenberg group is closely bound up with complex analysis in the unit ball. Here we present an exhaustive classification of affine-invariant function algebras and function spaces on the Heisenberg group, together with an example of an application to complex analysis in the unit ball. Contrary to the case of semisimple Lie groups, the class of invariant algebras on the Heisenberg group is far from trivial; it contains the algebras of boundary values of holomorphic functions (defined in a Siegel domain, which is naturally connected to the Heisenberg group), which play a special role in the classification. Transition to the complex sphere leads to a classification of Möbius spaces and algebras on the sphere. This classification includes, in particular, the results of Nagel and Rudin [84].

In Chapter III we study biholomorphically invariant (Möbius) spaces of functions defined on hermitian symmetric spaces, with the latter represented as bounded symmetric domains in $\mathbb{C}^n$. The central result of the chapter is the above-mentioned classification of Möbius spaces and algebras on the Shilov boundaries of bounded symmetric domains. We obtain detailed information about Möbius function spaces in the unit disk.

Chapter IV applies our results concerning invariant spaces to complex analysis in symmetric domains. Symmetric domains, both particular cases of them and generalizations (such as classical domains, the generalized disk, tubular domains, and Siegel domains) play an important role in complex analysis. Results established for symmetric domains are often prototypes of theorems for broader classes of domains. We will present characterizations (which are sometimes new even in the single-variable case) of holomorphic functions and their boundary values, using the homogeneous structure of symmetric domains.

Similar results are proved for other classes of domains as well.

E. A. Gorin inspired me to start working on the problems reflected in this book. I am extremely grateful to him for this.

It is my pleasure to express my gratitude to A. B. Alexandrov, S. G. Gindikin, E. A. Gorin, G. M. Henkin, V. P. Khavin, A. A. Kirillov, S. L. Krushkal, N. K. Nikolskiĭ, and Yu. G. Reshetnyak for numerous stimulating discussions of the results presented in this book.

CHAPTER I

Function Spaces and Function Algebras on Differentiable Manifolds and Symmetric Spaces of Noncompact Type

§1. Maximal ideal spaces of antisymmetric algebras on closed differentiable manifolds

Let X be a compact topological space and $C(X)$ the algebra (with respect to pointwise multiplication) of complex-valued functions on X, endowed with the uniform norm.

A subalgebra $A \subseteq C(A)$ is called antisymmetric if, whenever $f \in A$ and $\overline{f} \in A$, then f is a constant.

In this section we discuss closed antisymmetric subalgebras of $C(M)$ on a real differentiable manifold M. The main result is that, if M has no boundary, then there are no antisymmetric algebras of differentiable functions on M whose maximal ideal space is M.

We denote the maximal ideal space of an algebra A by $\mathfrak{M}(A)$.

1.1. Lemma. *Let M be a closed k-dimensional C^r-manifold, $r \geq \frac{k}{2}$. Let A be a closed antisymmetric subalgebra $C(M)$ such that $\mathfrak{M}(A) = M$. Then, for any set of functions $f_1, \dots, f_m \in A$ that separates points of M and any open subset $U \subset M$ such that $f_1, \dots, f_m \in C^r(U)$,*

$$\operatorname{rank}\left(\frac{\partial f_i}{\partial t_j}(x)\right)_{i,j=1}^{m,k} < k,$$

where $t_1, \dots, t_k$ are local coordinates on M and $x \in U$ is arbitrary.

Proof. Suppose the contrary, i.e., there exist $f_1, \dots, f_m \in A$ and an open set $U \subset M$ such that $f_i \in C^r(U)$ and

$$\operatorname{rank}\left(\frac{\partial f_i}{\partial t_j}(x)\right)_{i,j=1}^{m,k} = k, \qquad x \in U. \tag{1}$$

Then the map

$$\sigma(f_1, \dots, f_m)(x) = (f_1(x), \dots, f_m(x))$$

is regular on U and the image $V = \sigma(f_1, \ldots, f_m)(U)$ is a real (k-dimensional) C-submanifold of $\mathbb{C}^m$.

It follows from (1) that V is a totally real submanifold of $\mathbb{C}^m$, i.e., the tangent space at any point of V does not contain complex lines. We now apply the theorem of Hörmander and Wermer [16] about approximation on totally real submanifolds: any function $\chi \in C(\sigma(f_1, \ldots, f_m)(M))$ that vanishes on $\sigma(f_1, \ldots, f_m)(M)\backslash V$ can be approximated uniformly on $\sigma(f_1, \ldots, f_m)(M)$ by a sequence of functions χ_m, each of which is analytic in a neighborhood of the compact $\sigma(f_1, \ldots, f_m)(M)$.

Since $\mathfrak{M}(A) = M$, the image $\sigma(f_1, \ldots, f_m)(M)$ is the common spectrum of the elements $f_1, \ldots, f_m$ of A. According to the Shilov-Arens-Calderon theorem [19], the functions $g_m = \chi_m \circ \sigma(f_1, \ldots, f_m)$ belong to A. Since g_m converges uniformly to $\chi \circ \sigma(f_1, \ldots, f_m)$, the function $\chi \circ \sigma(f_1, \ldots, f_m)$ is also in A. However, χ may be chosen real and not a constant, and this contradicts the antisymmetric property of A.

1.2. Let N be a k-dimensional real differentiable submanifold of $\mathbb{C}^n$. Denote the tangent space of N at a point $p \in N$ by $T_p(N)$ and the complex tangent space $T_p(N) \cap iT_p(N)$ by $H_p(N)$.

Recall that N is a CR-space if the dimension of $H_p(N)$ is the same for all $p \in N$. For any CR-manifold N we can define the holomorphic tangent bundle with the fiber $H_p(N)$.

We will need an extension theorem for CR-manifolds due to Hunt and Wells [75].

THEOREM (HUNT-WELLS). *Let N be a real k-dimensional CR-submanifold of an open set in $\mathbb{C}^n$, and suppose that the Levi form on N does not vanish at $p \in N$. Then there exists a $(k+1)$-dimensional CR-manifold $\widetilde{N}$ such that in the neighborhood of p the boundary of $\widetilde{N}$ contains N, and every CR-function f on m N has an extension $\widetilde{f}$ to* int $\widetilde{N}$ *such that $\widetilde{f}$ is a CR-function on $\widetilde{N}$.*

The manifold $\widetilde{N}$ is constructed by Bishop's technique of gluing analytic disks to N.

We now prove a key lemma for the main result of the section. We first need a definition.

DEFINITION. A set of functions $B \subset C^1(M)$ on a differentiable manifold M is *locally regular* if for any point $p \in M$ one can find $f_1, \ldots, f_m \in B$ such that

$$\operatorname{rank}_{\mathbb{R}} \left(\frac{\partial g_i}{\partial t_j}(x) \right)_{i,j=1}^{2m,k} = k,$$

$$g_i = \operatorname{Re}(f_i), \qquad g_{i+m} = \operatorname{Im}(f_i), \qquad i = 1, \ldots, m.$$

LEMMA. *Let M be a k-dimensional closed C^∞-manifold and A be a closed antisymmetric subalgebra of $C(M)$ such that $\mathfrak{M}(A) = M$. Suppose* *that the set* $A \cap C^\infty$ *is locally regular and separates points of* M. *Then there exist an open set* $U \subset M$ *and functions* $f_1, \dots, f_m \in A \cap C^\infty$ *such that*:

(1) *the map* $\sigma(f_1, \dots, f_m)(x) = (f_1(x), \dots, f_m(x))$ *is an isomorphism of* U *onto the* CR*-manifold* $N = \sigma(f_1, \dots, f_m)(U)$;

(2) *the Levi form of* N *is nondegenerate at some point*;

(3) *for any* $h \in A \cap C^\infty(M)$, *the function*

$$h \circ [\sigma(f_1, \dots, f_m)]^{-1}$$

is CR *on* N.

PROOF. Since M is compact and $A \cap C^\infty(M)$ is locally regular, there exists a finite set of functions $f_1, \dots, f_i \in A \cap C^\infty(M)$ which separates points of M. For the same reason, we can add functions $f_{i+1}, \dots, f_l \in A \cap C^\infty(M)$ in such a way that the map $\sigma(f_1, \dots, f_l) : M \to \mathbb{C}^l$ is a be C^∞-regular embedding. Set $S = \sigma(f_1, \dots, f_l)(M)$.

By a theorem of Wells [48], the compact manifold S contains a nonempty open subset $W \subset S$, each point of which is a peak point for the algebra $\mathscr{O}(S)$ of functions holomorphic in a neighborhood of S.

Set $V = [\sigma(f_1, \dots, f_l)]^{-1}(W)$. Let $\mathscr{R}$ be the set of all finite sets $\{h_1, \dots, h_s\} \subset A \cap C^\infty(M)$ that separate points on M.

Further, set

$$r(h_1, \dots, h_s) = \max_{x \in V} \operatorname{rank}_{\mathbb{C}} \left(\frac{\partial h_i}{\partial t_j}(x) \right)_{i,j=1}^{s,k},$$

and let $f_{l+1}, \dots, f_m \in A \cap C^\infty(M)$ be a set of functions on which maximum

$$r(f_{l+1}, \dots, f_m) = \max_{\{h_1, \dots, h_s\} \in \mathscr{R}} r(h_1, \dots, h_s)$$

is attained. Let $x_0 \in V$ be a point such that

$$\operatorname{rank}_{\mathbb{C}} \left(\frac{\partial f_i}{\partial t_j}(x_0) \right)_{i,j=l+1,1}^{m,k} = \max_{x \in V} \operatorname{rank}_{\mathbb{C}} \left(\frac{\partial f_i}{\partial t_j}(x) \right)_{i,j=l+1,1}^{m,k}$$

As the rank is maximal at x_0, it is constant in a neighborhood $U \subset V$ of x_0.

Now consider the map

$$\sigma(f_1, \dots, f_m) : U \to \mathbb{C}^m,$$

which is of course a C^∞-regular embedding. Set

$$N = \sigma(f_1, \dots, f_m)(U).$$

We can illustrate these constructions by a commutative diagram

$$\begin{array}{ccc} U & \xrightarrow{\quad \sigma^m \quad} & N \\ \| & & \downarrow \pi \\ U & \xrightarrow{\sigma^l} & \sigma(f_1, \dots, f_l)(U) \subset W \subset S \end{array}$$

where $\sigma^j = \sigma(f_1, \dots, f_j)$ and π is the orthogonal projection of $\mathbb{C}^m$ onto $\mathbb{C}^n$, $\pi(z) = (z_1, \dots, z_l)$.

We claim that N is a CR-manifold with a nondegenerate complex tangent bundle. Indeed, let $p \in N$ and $p = \sigma(f_1, \dots, f_m)(x)$ for some $x \in U$. We have

$$k - \dim_{\mathbb{C}} H_p(N) = \operatorname{rank}_{\mathbb{C}} \left(\frac{\partial f_i}{\partial t_j}(x) \right)_{i,j=1}^{m,k}$$
$$\geq \operatorname{rank}_{\mathbb{C}} \left(\frac{\partial f_i}{\partial t_j}(x) \right)_{i,j=l+1,1}^{m,k}.$$

But $f_{l+1}, \dots, f_m$ and U were chosen in such a way that strict inequality is impossible, and besides, the quantity on the right is the same for all $x \in U$. Therefore,

$$\dim_{\mathbb{C}} H_p(N) = k - \operatorname{rank}_{\mathbb{C}} \left(\frac{\partial f_i}{\partial t_j}(x) \right)_{i,j=1}^{m,k} = d,$$

where d is independent of $x \in U$. By Lemma 1.1, $d > 0$.

We now prove that the Levi form $L_q(N)$ does not vanish for all $q \in N$. Projection $\pi : \mathbb{C}^m \to \mathbb{C}^l$ is a diffeomorphism of N onto $\sigma(f_1, \dots, f_l)(U)$ and induces a map of complex tangent spaces

$$d\pi_q : H_q(N) \to H_{\pi(q)}(\pi(N)).$$

This map coincides with the projection $d\pi(w) = \pi(w)$, $w \in H_q(N)$. The projection of the Levi form $L_q(N)$ is the Levi form:

$$L_q(N)(w) = L_{\pi(q)}(\pi(N))\ (d\pi_q(w)).$$

Suppose that $L_q(N) = 0$ for all $q \in N$. Then $L_p(\pi(N)) = 0$ for all $p \in \pi(N)$. By Lemma 1.1, the manifold $\pi(N) = \sigma(f_1, \dots, f_l)(U)$ has a nonzero complex tangent space $H_p(\pi(N))$ at each point. Thus there exists an open set $U' \subset \pi(N)$ at whose points the dimension of these spaces is maximal and constant. Hence U' is a CR-manifold with trivial Levi form and nontrivial complex tangent space. By a theorem of Zommer and Nirenberg (see [47]), U' is a foliation whose leaves are holomorphic curves, but this is impossible, since $U' \subset W$, and by choice of the neighborhood $W \subset S$ all the points of W are peak points of $\mathscr{O}(S)$.

Thus the first two assertions of the theorem are true. In order to prove the third, choose an arbitrary element $h \in A \cap C^\infty(M)$ and consider the map

$$F_h = \sigma(f_1, \dots, f_m, h) : \ U \to \mathbb{C}^{m+1}.$$

Let λ denote the orthogonal projection onto the first m coordinates, $\lambda : \mathbb{C}^{m+1} \to \mathbb{C}^m$. Then λ is a diffeomorphism of $F_h(U)$ onto N.

Choose CR-vector fields

$$X_1, \dots, X_d, \qquad d = \dim_{\mathbb{C}} H(F_h(U)),$$

that form a basis in the complex tangent bundle $H(F_h(U))$, and set $Y_i = d\lambda(X_i)$, $i = 1, \dots, d$. We claim show that $Y_1, \dots, Y_d$ form a basis in $H(N)$. Indeed,

$$\operatorname{rank}_{\mathbb{C}} \left(\frac{\partial f_i}{\partial t_j}(x)\right)_{i,j=1}^{m+1,k} \geq \operatorname{rank}_{\mathbb{C}} \left(\frac{\partial f_i}{\partial t_j}(x)\right)_{i,j=1}^{m,k}$$

$$\geq \operatorname{rank}_{\mathbb{C}} \left(\frac{\partial f_i}{\partial t_j}(x)\right)_{i,j=l+1,1}^{m,k}, \qquad f_{m+1} = h,$$

for all $x \in U$. But by the construction of $f_{l+1}, \dots, f_m$ and the choice of U, the first and the third ranks coincide; hence,

$$\operatorname{rank}_{\mathbb{C}} \left(\frac{\partial f_i}{\partial t_j}(x)\right)_{i,j=1}^{m+1,k} = \operatorname{rank}_{\mathbb{C}} \left(\frac{\partial f_i}{\partial t_j}(x)\right)_{i,j=1}^{m,k}, \qquad x \in U.$$

This means that the defects of the spaces $H_q(F_h(U)$ and $H_{\lambda(q)}(N)$ for $q = F_h(x)$ are equal, and therefore

$$\dim_{\mathbb{C}} H_{\lambda(q)}(N) = \dim_{\mathbb{C}} H_q(F_h(U)) = d.$$

Since $d\lambda$ is an isomorphism of tangent spaces, it follows that $Y_1, \dots, Y_d$ are linearly independent at every point of N, so they form a basis of $H(N)$.

Now we see that

$$\overline{Y}_i \left\{h \circ [\sigma(f_1, \dots, f_m)]^{-1}\right\} = \overline{X}_i \left\{h \circ [\sigma(f_1, \dots, f_m)]^{-1} \circ \lambda\right\},$$

where $\overline{X}_i$, $\overline{Y}_i$ are appropriate CR-antiholomorphic tangent vector fields. Since

$$\left\{h \circ [\sigma(f_1, \dots, f_m)]^{-1} \circ \lambda\right\}(z_1, \dots, z_{m+1}) = z_{m+1}$$

on $F_h(U)$, we see that

$$\overline{X}_i \left\{h \circ [\sigma(f_1, \dots, f_m)]^{-1} \circ \lambda\right\} = \overline{X}_i(z_{m+1}) = 0;$$

therefore, the function $h \circ [\sigma(f_1, \dots, f_m)]^{-1}$ is annihilated by the vector fields $\overline{Y}_i$, $i = 1, \dots, d$. Hence it satisfies the Cauchy-Riemann tangent equations on N, and this completes the proof.

1.3. THEOREM. *Let M a be closed k-dimensional real C^∞-manifold and A a closed antisymmetric subalgebra of $C(M)$. Suppose that $A \cap C^\infty(M)$ is dense in A and locally regular on M. Then $\mathfrak{M}(A) \neq M$.*

PROOF. Choose functions $f_1, \dots, f_m \in A \cap C^\infty(M)$, a neighborhood U, and a manifold N as in Lemma 1.1. Since N satisfies the conditions of the Hunt-Wells theorem, there exists a holomorphic map $j : B^1 \to \mathbb{C}^m$ of the complex unit disk B^1 such that $j(\partial B^1) \subset N$, and every CR-function on N has an analytic continuation to the analytic disk $j(B^1)$.

Let $h \in A \cap C^\infty(M)$. By Lemma 1.1, $h \circ [\sigma(f_1, \dots, f_m)]^{-1}$ is a CR-function on N; hence, it may be extended in $j(B^1)$ as an analytic function $\widetilde{h}$. For $z \in B^1$, set

$$m_z(h) = \widetilde{h}(j(z)).$$

The functional m_z is linear and multiplicative, and by the maximum principle

$$|m_z(h)| \leq \sup_{w \in \partial B^1} |\widetilde{h}(j(w))| \leq \sup_{x \in M} |h(x)|.$$

This implies that m_z may be extended from the dense subset $A \cap C^\infty(M)$ to a homomorphism $m_z : A \to \mathbb{C}$. Choosing $z \in B^1$ so that $j(z) \notin M$, we obtain a functional belonging to $\mathfrak{M}(A)$, that corresponds to no point of M. This proves the theorem.

1.4. We now proceed to the two-dimensional case. We have already pointed out (Introduction) that the first example of an antisymmetric algebra on a closed two-dimensional manifold whose maximal ideal space is the manifold itself was the Arens algebra $A(S^2, K)$, i.e., the set of all functions continuous on the Riemann sphere S^2 and analytic everywhere except a fixed nowhere dense set K of positive Lebesgue measure. Arens's idea was developed by Brandstein [59]. His starting point was the algebra of functions analytic in a disk. Using the fact that any compact two-dimensional manifold can be derived from the unit disc by "gluing together" some arcs of the disk boundary, he constructed antisymmetric algebras on all arbitrary two-dimensional manifolds.

The following theorem shows that the Arens-Brandstein construction is in fact the only possible one and that all antisymmetric algebras on closed two-dimensional manifolds are in some way similar to the Arens algebra.

THEOREM. *Let M be a closed two-dimensional C^2-manifold and A a closed antisymmetric subalgebra of $C(M)$ such that $\mathfrak{M}(A) = M$. Let B be a dense subset of A and U an open subset of M such that $B|_U \subset C^2(U)$ and B is locally regular on U. Then one can define a complex manifold structure on U such that all elements of A become analytic functions.*

PROOF. Pick $p \in U$. There exist $f_1, \dots, f_m \in B$ such that the map $\sigma(f_1, \dots, f_m)$ is regular in a neighborhood of p. Without loss of generality, we may assume that the first four columns of Jacobian contain a minor which does not vanish in some neighborhood $U_p \subset U$ of p:

$$\operatorname{rank}_{\mathbb{R}} \begin{pmatrix} \dfrac{\partial \operatorname{Re} f_1}{\partial t_1} & \dfrac{\partial \operatorname{Im} f_1}{\partial t_1} & \dfrac{\partial \operatorname{Re} f_2}{\partial t_1} & \dfrac{\partial \operatorname{Im} f_2}{\partial t_1} \\ \dfrac{\partial \operatorname{Re} f_1}{\partial t_2} & \dfrac{\partial \operatorname{Im} f_1}{\partial t_2} & \dfrac{\partial \operatorname{Re} f_2}{\partial t_2} & \dfrac{\partial \operatorname{Im} f_2}{\partial t_2} \end{pmatrix} (x) = 2, \qquad x \in U_p.$$

On the other hand, by Lemma 1.1,

$$\operatorname{rank}_{\mathbb{C}} \begin{pmatrix} \dfrac{\partial f_1}{\partial t_1} & \dfrac{\partial f_2}{\partial t_1} \\ \dfrac{\partial f_1}{\partial t_2} & \dfrac{\partial f_2}{\partial t_2} \end{pmatrix} (x) < 2, \qquad x \in U_p,$$

and therefore

$$\operatorname{rank}_{\mathbb{C}} \begin{pmatrix} \dfrac{\partial f_1}{\partial t_1} & \dfrac{\partial f_2}{\partial t_1} \\ \dfrac{\partial f_1}{\partial t_2} & \dfrac{\partial f_2}{\partial t_2} \end{pmatrix} (x) = 1, \qquad x \in U_p.$$

This means that for every $q \in \sigma(f_1, f_2)(U)$ the complex tangent space $H_q(N)$ of manifold $N = \sigma(f_1, f_2)(U_p) \subset \mathbb{C}^2$ has dimension one over $\mathbb{C}$ and coincides with the real tangent space $T_q(N)$.

Let $q = \sigma(f_1, f_2)(p)$. Choose the neighborhood U_p small enough, so that the orthogonal projection of the neighborhood $V_p = \sigma(f_1, f_2)(U_p)$ of q onto $H_q(N)$ is one-to-one. Call this projection π_p, and consider the maps

$$\alpha_p = [\sigma(f_1, f_2)]^{-1} \circ \pi_p^{-1} : \pi_p(V_p) \to M.$$

We will show that the pairs (α_p, V_p), $p \in U$, define the desired complex structure on U. By Lemma 1.1, for $f \in B$ we have

$$d f \wedge d f_1|_U = d f \wedge d f_2|_U = 0.$$

Substituting $\varphi = [\sigma(f_1, f_2)]^{-1} : V_p \to U_p$, we get

$$\varphi^* d f_1 = d z_1, \qquad \varphi^* d f_2 = d z_2.$$

Hence

$$d(f \circ \varphi) \wedge d z_i = \varphi^*(d f \wedge d f_i) = 0, \qquad i = 1, 2,$$

implying that $f \circ \alpha_p$ is analytic on $\pi_p(V_p)$. Suppose that $U_1 \cap U_2 \neq \varnothing$. By construction, α_{p_1}, α_{p_2} may be written as

$$\alpha_{p_i} = \left[\sigma(f'_{p_i}, f''_{p_i})\right]^{-1} \circ \pi_{p_i}^{-1}, \qquad i = 1, 2,$$

for some $f'_{p_i}, f''_{p_i} \in B$. We have already shown that $f'_{p_1} \circ \alpha_{p_2}$, $f''_{p_1} \circ \alpha_{p_2}$ are analytic on $\pi_{p_2}(V_{p_2})$. Since

$$\alpha_{p_1}^{-1} \circ \alpha_{p_2} = \pi_{p_1}\left(\sigma(f'_{p_1} \circ \alpha_{p_2}, f''_{p_1} \circ \alpha_{p_2})\right),$$

we conclude that the transition function $\alpha_{p_1}^{-1} \circ \alpha_{p_2}$ is analytic in the coordinates that we have defined on $U_{p_1} \cap U_{p_2}$. We have thus defined a complex structure on U such that all functions are analytic in B (hence, also in A, since B is dense in A).

REMARK. Most probably, there are antisymmetric algebras on closed real differentiable manifolds of any dimension. On the grounds of the results of the present section one can conjecture that these algebras are constructed as follows: outside some closed subset—call it the singular subset—the manifolds are foliations such that one can introduce complex structure on the leaves so that the elements of the algebra are just the analytic functions. The closures of these leaves have common points in the singular subset, and this guarantees that the algebra is antisymmetric.

We illustrate this conjecture with an example of an antisymmetric algebra on the n-sphere S^n.

Set $X = S^2 \times D^{n-2}$, where D^{n-2} is the closed unit ball in $\mathbb{R}^{n-2}$. Let $K \subset S^2$ be a closed nowhere dense subset of S^2 of positive measure and $A(S^2, K)$ the Arens algebra as defined above.

Let A denote the algebra of functions on X such that

(1) $f|_{S^2\times\{t\}} \in A(S^2, K)$ for all $t \in D^{n-2}$;

(2) $f|_{S^2\times\{t\}} = \text{const}$ for all $t \in \partial D^{n-2}$;

(3) $f|_{\{s_0\}\times D^{n-2}} = \text{const}$ for some previously fixed $s_0 \in K$.

Using the isomorphism $\mathfrak{M}(A(S^2, K)) \simeq S^2$, one can show that the maximal ideal space of A may be constructed from the "cylinder" X by gluing the sets $S^2 \times \{t\}$ to a point, where $t \in \partial D^{n-2}$, and also gluing $\{s_0\} \times D^{n-2}$ to a point. The space thus obtained is homeomorphic to the n-sphere. That the algebra A (considered as a subalgebra of $C(S^n)$) is antisymmetric follows from the fact that $A(S^2, K)$ is antisymmetric and the third condition.

This example illustrates the construction we have sketched above. The singular set is the subset obtained from $K \times D^{n-2}$ by gluing. In an analogous way one can construct antisymmetric algebras on spheres with handles. It may be possible to construct an antisymmetric algebra on any compact manifold starting from the antisymmetric algebra on the sphere S^n, by using the spherical surgery and the technnique of [59].

§2. Algebras on manifolds invariant under diffeomorphism groups

In this section we proceed to study invariant algebras. The antisymmetry property, so crucial in the preceding section, will now follow from the invariance of the algebra. In addition, the existence of a transitive group acting on the manifold will imply the existence of a dense subset of differentiable functions in the invariant algebra of the group. This will allow us to apply the preceding results for a description of invariant algebras.

2.1. We first recall a notion from the theory of uniform algebras. Let M be a compact set and A a closed subalgebra of $C(M)$. A subset K is called *the antisymmetry set* of A if every function in A that is real on K is constant on K. The maximal antisymmetry sets (with respect to inclusion) form a partition of M into closed disjoint subsets. The importance of this construction is due to the following theorem of Bishop (see [19]): if $f \in C(M)$ and the restriction $f|_K$ to any maximal antisymmetry set can be extended to M as an element in A, then $f \in A$.

2.2. THEOREM. *Let M be a closed real C^∞-manifold and G a Lie group of diffeomorphisms of M acting transitively on M. Suppose that every G-homogeneous partition of M has a discrete fiber. Then $C(M)$ is the only closed invariant algebra of continuous functions on M such that the maximal ideal space is M.*

The proof uses the following simple assertion.

2.3. LEMMA. *Let $\mathscr{F}$ be a set of real C^1 functions that separate the points of M. Then the set of $x \in M$ such that the $\operatorname{rank} d_x f = \dim M$ for some $f = (f_1, \dots, f_m)$, $f_1, \dots, f_m \in \mathscr{F}$, is dense in M. Here $d_x f$ is the Jacobian at the point x.*

PROOF. Let U be an arbitrary compact parametrized neighborhood in M. Set

$$r = \max_{\substack{f_i \in \mathscr{F} \\ m=1,2,\dots}} \operatorname{rank} \left(\frac{\partial f_i}{\partial t_j}(x) \right)_{i,j=1}^{m,k},$$

where $t_1, \dots, t_k$ are local coordinates in U, $k = \dim M$.

Let us take a collection $f_1, \dots, f_m$, $f_i \in \mathscr{F}$, and a point $x \in U$ for which this maximum is attained. Suppose that $r < k$. Since the rank of Jacobian

$$\left(\frac{\partial f_i}{\partial t_j}(x) \right)_{i,j=1}^{m,k}$$

is constant in a neighborhood of X, we may assume that the coordinates t_j have been chosen in such a way that $f_i(t) = t_i$, $i = 1, \dots, r$, in a neighborhood $V \subset U$ of x. Consider the fiber $s = \{t_1 = a_1, \dots, t_r = a_r\}$ through V. Since $\mathscr{F}$ separates points, there exists $h \in \mathscr{F}$ that is not a constant on s. But then at some point of s, the rank of

$$\left(\frac{\partial f_i}{\partial t_j}(x) \right)_{i,j=1}^{m+1,k}, \qquad f_{m+1} = h,$$

is greater than r, and this is a contradiction with the definition of r. Thus $r = k$.

PROOF OF THEOREM 2.2. Let A be a G-invariant closed subalgebra of $C(M)$ such that $\mathfrak{M}(A) = M$. Let F be a maximal antisymmetry set (MAS for short) of A. By invariance, the partition of M into MAS is G-homogeneous, and by assumption either F is discrete or $F = M$. Since A separates the points of F, in the first case we have $A|_F = C(F)$, and by the Bishop's theorem $A = C(M)$. In the other case we have $F = M$ and A is antisymmetric. Let us study this situation in more detail.

Consider a δ-sequence $\chi_n \in C^\infty(G)$ of positive functions with compact support. Denote Haar measure on G by dg, and fix a point $o \in M$.

For any $f \in A$, the functions

$$f_n(x) = \frac{1}{\int_G \chi_n(g)dg} \int_G \chi_n(g) f(g^{-1}(x))\, dg$$

belong to A and converge to f uniformly as $n \to \infty$. On the other hand, after a change of variables we get

$$f_n(x) = \frac{1}{\int_G \chi_n(g)dg} \int_G \chi_n(g_x h^{-1}) f(h(o))\, dh,$$

where g_x is an element of G such that $g_x(o) = x$.

Let K denote the stationary subgroup of G at o. Then M may be regarded as the space of left cosets $M = G/K$. Being closed, K is a Lie subgroup. Let $\mathfrak{G}$ be the Lie algebra of G and $\mathfrak{N}$ a complementary vector subspace of the Lie algebra $\mathfrak{K}$ of K: $\mathfrak{G} = \mathfrak{K} \oplus \mathfrak{N}$. There exists a neighborhood V of zero in $\mathfrak{N}$ such that the map $\psi = \pi \circ \exp$ from $\mathfrak{G}$ to M, where π is a canonical projection of G onto M, $\pi(g) = g(o)$, is a C^∞ diffeomorphism of V to a neighborhood, say W, of o in M (see [28, Section 4]).

The inverse of ψ is a C^∞-map which associates to each $x \in W$ an element $g_x \in G$, $g_x(o) = x$. The second expression for f_n now implies that $f_n \in C^\infty(M)$. Thus $A \cap C^\infty(M)$ is dense in A.

The set $A \cap C^\infty(M)$ is locally regular. Indeed, since $\mathfrak{M}(A) = M$ and $A \cap C^\infty(M)$ is dense in A, it follows that $A \cap C^\infty(M)$ separates points on M. By Lemma 2.3, there exists a point $x \in M$ such that

$$\operatorname{rank} \sigma(\operatorname{Re} f_1, \operatorname{Im} f_1, \ldots, \operatorname{Re} f_m, \operatorname{Im} f_m) = k$$

for some $f_1, \ldots, f_m \in A \cap C^\infty(M)$. It follows from the invariance of A and the transitivity of the action of G on M that there exist such functions for every point of M. We can therefore apply Theorem 1.3 so that $\mathfrak{M}(A) \neq M$.

REMARK. The assumptions of Theorem 2.2 are satisfied if Lie algebra $\mathfrak{K}$ of the isotopy subgroup K is a maximal subalgebra of the Lie algebra $\mathfrak{G}$ of G. Indeed, let S be a G-homogeneous partition of M. Let o be a fixed point of K and $s_o \in S$ the fiber containing o. Let H be the subgroup of G consisting of all g such that $gs_o \subset s_o$. Then H is a closed subgroup and, therefore, a Lie group. Since $K \subset H$, we see that

$$\mathfrak{K} \subset \mathfrak{H} \subset \mathfrak{G}.$$

Since $\mathfrak{K}$ is maximal, either $\mathfrak{H} = \mathfrak{G}$, or $\mathfrak{H} = \mathfrak{G}$. In the first case we have $H = G$, i.e., $s_o = M$. In the second case, $s_o = H/K$ is discrete.

§3. Invariant spaces and algebras of integrable functions on symmetric spaces of noncompact type

In this section we derive necessary and sufficient conditions on symmetric spaces of noncompact type under which any closed isometry-invariant space of square-integrable functions is closed under complex conjugation. The results will be applied to the description of invariant algebras.

3.1. We shall first formulate a simple statement that will be used later in various situations.

LEMMA. *Let X be the homogeneous space of a locally compact group G and $C_0(X)$ be the space of all complex-valued continuous functions on X that vanish at infinity. Let Y be a G-invariant closed subspace of $C_0(X)$. Then*

$$C_0(X) \cap \mathrm{cl}_{L^p}\left[Y \cap L^p\right] \subset Y,$$

where $L^p = L^p(X, dx)$, $p = 1, 2, \dots$, and dx is a G-invariant measure on X.

PROOF. Note that the space $Y^{\perp}$ of all Radon measures orthogonal to Y contains a dense (in weak topology) subset of measures that are absolutely continuous with respect to dx and have bounded Radon-Nikodym derivatives. This can be proved by considering the convolutions

$$\mu_{\varphi} = \int_G \varphi(g)(\mu \circ g)dg$$

of measures μ with compactly supported continuous functions φ on G. These convolutions form a weakly dense subset of $Y^{\perp}$, but each measure μ_{φ} is absolutely continuous and $d\mu_{\varphi}/dx$ is a bounded function.

Let $p > 1$. If μ is orthogonal to Y, then for every $\varphi \in C_c(G)$ the measure μ_{φ} is orthogonal to $\mathrm{cl}_{L^p}[Y \cap L^p]$, because $\mu_{\varphi} = h\,dx$ for $h \in L^q$, $1/p + 1/q = 1$. Since the measures μ_{φ} approximate μ in the weak topology, it follows that μ is orthogonal to

$$C_0(X) \cap \mathrm{cl}_{L^p}\left[Y \cap L^p\right].$$

By the Hahn-Banach theorem, this proves the inclusion.

3.2. Let X be a connected irreducible symmetric Riemann space of noncompact type, $X = G/K$ (the space of left cosets), where G is a connected semisimple noncompact Lie group with finite center and K the maximal compact subgroup.

Let T_g denote the left translation operator on X:

$$(T_g f)(x) = f(g^{-1}x), \qquad x \in G.$$

Let $L^p_K(X)$ denote the subspace in $L^p(X, dx)$ which consists of the K-invariant functions, i.e., f such that $f(kx) = f(x)$, $k \in K$, where dx is a G-invariance measure on X.

Consider the K-spherical Fourier transform on X:

$$\hat{f}(\lambda) = \int_X f(x)\overline{\varphi_{\lambda}(x)}\,dx,$$

where φ_λ is a bounded spherical function corresponding to a vector λ in a Cartan (i.e., maximal abelian) subalgebra $\mathfrak{H} \subset \mathfrak{G}$.

The Plancherel measure on $\mathfrak{H}$ will be denoted by $d\lambda$. The Fourier transform can be extended from $L^1_K(X) \cap L^2_K(X)$ to an isometry of $L^2_K(X)$ and of the space $L^2_W(\mathfrak{H}, d\lambda)$ of functions invariant under the Weyl group W.

Recall that the Weyl group is the (finite) quotient group

$$W = N(\mathfrak{H})/K^{\mathfrak{K}}$$

of the normalizer

$$N(\mathfrak{H}) = \{k \in K : \mathrm{Ad}(k)(\mathfrak{H}) = \mathfrak{H}\}$$

of the Cartan subalgebra $\mathfrak{H}$ of K by the centralizer

$$K^{\mathfrak{H}} = \{k \in K : \mathrm{Ad}(k)(\lambda) = \lambda, \ \lambda \in \mathfrak{H}\}.$$

Let ξ be a λ-measurable W-symmetric subset of $\mathfrak{H}$. Let $L^2_{K,\xi}(X)$ be the subspace of $L^2_K(X)$ consisting of all f such that $\hat{f}(\lambda) = 0$ almost everywhere outside ξ.

Define the orthogonal projection $S_K : L^2(X) \to L^2_K(X)$ by

$$S_K f = \int_K (T_k f)\, dk.$$

Lemma. *Let Y be a closed subspace of $L^2_K(X)$ invariant under all operators $S_K T_g$, $g \in G$. Then $Y = L^2_{K,\xi}(X)$ for some $\xi \subset \mathfrak{H}/W$.*

Proof. Denote the natural projection of G onto X by π, and let $o = \pi(e)$.

For an arbitrary $g \in G$, consider the Fourier transform of $S_K T_g f$ for $f \in Y$:

$$\begin{aligned}\widehat{S_K T_g f}(\lambda) &= \int_K \int_X f(g^{-1}kx)\overline{\varphi_\lambda(x)}\, dx = \int_X f(x)\overline{\varphi_\lambda(gx)}\, dx \\ &= \int_G f(h(o))\overline{\varphi_\lambda(gh(o))}\, dh = \int_K \int_G f(h(o))\overline{\varphi_\lambda(gkh(o))}\, dg\, dk.\end{aligned}$$

The spherical functions $\widetilde{\varphi}_\lambda(g) = \varphi_\lambda(g(o))$ on G satisfy the functional equation

$$\int_K \widetilde{\varphi}_\lambda(gkh)\, dk = \widetilde{\varphi}_\lambda(g) \cdot \widetilde{\varphi}_\lambda(h),$$

and hence

$$\widehat{S_K T_g f}(\lambda) = \overline{\widetilde{\varphi}_\lambda(g)} \cdot \hat{f}(\lambda).$$

We conclude that the set $\widehat{Y}$ of the Fourier transforms of functions of Y is invariant under multiplication by functions $\lambda \to \overline{\widetilde{\varphi}_\lambda(g)}$, $g \in G$. Therefore, it is also invariant under the multiplication by $\hat{h}$, $h \in L^1_K(X)$. The set of these functions is dense in $L^\infty_W(\mathfrak{H})$, so $\widehat{Y}$ is invariant under multiplication by any bounded measurable function. Thus $\widehat{Y}$ must have the form $\chi_\xi L^2_W(\mathfrak{H})$, for some measurable subset ξ of $\mathfrak{H}/W$. This proves the lemma.

Define an involution operation on functions defined on G by $f^\#(g) = \overline{f(g^{-1})}$. If f is defined on X and is K-invariant, then one can define an involution

$$f^\# = (\pi^*)^{-1}(\pi^* f)^\#,$$

where $\pi^* f = f \circ \pi$ is the map induced in the function spaces by the natural projection π.

COROLLARY. *The involution $f \to f^\#$ is defined in every closed subspace of $L^2_K(X)$ that is invariant under the operators $S_K T_g$, $g \in G$.*

PROOF. As the spherical functions $\widetilde{\varphi}_\lambda$, $\lambda \in \mathfrak{H}$, are positive definite, $\widetilde{\varphi}_\lambda(g^{-1}) = \overline{\widetilde{\varphi}_\lambda(g)}$. Since the group is unimodular,

$$\widehat{f^\#}(\lambda) = \int_G \overline{f(g^{-1}(o))}\overline{\widetilde{\varphi}_\lambda(g)}\, dg = \int_G \overline{f(h(o))\widetilde{\varphi}_\lambda(h^{-1})}\, dh = \overline{\widehat{f}(\lambda)}.$$

Thus it is clear that all the spaces $L^2_{K,\xi}(X)$ are invariant under the involution $f \to f^\#$.

3.3. PROPOSITION. *Every G-invariant closed subspace $Y \subset L^2(X)$ is the closed G-invariant linear span $[Y \cap L^2_K(X)]_G$ of $Y \cap L^2_K(X)$.*

PROOF. We claim that

$$\left[Y \cap L^2_K(X)\right]_G^\perp \cap Y = \{0\}.$$

Using convolutions, one easily sees that the subspace of continuous functions is dense in $[Y \cap L^2_K(X)]_G^\perp \cap Y$. Let f be an element of this subspace. For any $x \in X$, choose $g_x \in G$ such that $g_x(o) = x$. Then

$$\|S_K T_{g_x} f\|^2_{L^2(X)} = \left\langle S_K T_{g_x} f,\ S_K T_{g_x} f\right\rangle_{L^2(X)} = \left\langle f,\ T_{g_x^{-1}} S_K T_{g_x} f\right\rangle_{L^2(X)} = 0,$$

because $T_{g_x^{-1}} S_K T_{g_x} \in [Y \cap L^2_K(X)]_G$. Therefore, $f(x) = (S_K T_{g_x} f)(o) = 0$. Thus the orthogonal complement of $[Y \cap L^2_K(X)]_G$ in Y is trivial. Q.E.D.

3.4. PROPOSITION. *The following conditions are equivalent:*

(1) *Every closed G-invariant subspace of $L^2(X)$ is closed under complex conjugation.*

(2) *The spherical functions φ_λ, $\lambda \in \mathfrak{H}$, take real values.*

PROOF. (1) $\Rightarrow$ (2). Let $\lambda_0 \in \mathfrak{H}$, and let V be an arbitrary W-symmetric neighborhood of λ_0 in $\mathfrak{H}$. Set $Y = L^2_{K,V}(X)$. Since

$$\overline{\hat{f}(\lambda)} = \hat{f}(\lambda^*),$$

where $\varphi_{\lambda^*} = \overline{\varphi}_\lambda$, it follows that $\overline{Y} = L^2_{K, V^*}(X)$, where $V^* = \{\lambda^* : \lambda \in V\}$. By assumption, $Y = \overline{Y}$; hence, V coincides with V^* modulo a set of measure zero; but since both sets are open, we have $V = V^*$. Since we can choose V as small as desired, $\lambda_0 = \lambda_0^*$, i.e., φ_{λ_0} is real.

(2) $\Rightarrow$ (1). If all the functions φ_λ, $\lambda \in \mathfrak{H}$, are real, then $\hat{\bar{f}} = \hat{\hat{f}}$, so any space $L^2_{K,\xi}$ is invariant under complex conjugation. We can now use Proposition 3.2.

3.5. THEOREM. *Every closed G-invariant subspace of $L^2(X)$ is invariant under complex conjugation if and only if the Weyl group $W = W(G, K)$ contains the symmetry* $-\mathrm{Id}$.

PROOF. By Proposition 3.3 it is enough to show that condition $-\mathrm{Id} \in W$ holds if and only if all the functions φ_λ, $\lambda \in \mathfrak{H}$, are real. We use the Harish-Chandra representation [28] of K-spherical functions:

$$\widetilde{\varphi}_\lambda(g) = \int_K e^{\langle \rho + i\lambda, \gamma(gk)\rangle} dk, \tag{1}$$

where ρ is half the sum of the positive roots of the Lie algebra $\mathfrak{G}$, γ the map from G to $\mathfrak{H}$ defined by Iwasawa decomposition, and $\langle\ ,\ \rangle$ the Killing form.

Formula (1) implies that $\overline{\widetilde{\varphi}}_\lambda = \widetilde{\varphi}_{-\lambda}$.

Suppose now that $-\mathrm{Id} \in W$. As the parameterization of the spherical functions is unique up to the action of the Weyl group, $\widetilde{\varphi}_\lambda = \widetilde{\varphi}_{-\lambda}$; hence, $\varphi_\lambda = \overline{\varphi}_\lambda$.

Conversely, suppose that all the functions φ_λ, $\lambda \in \mathfrak{H}$, are real. Then $\widetilde{\varphi}_{-\lambda} = \overline{\widetilde{\varphi}}_\lambda = \widetilde{\varphi}_\lambda$, and this means that $-\lambda = w_\lambda \lambda$ for some $w_\lambda \in W$.

Let $\mathfrak{H}_w$, $w \in W$, be the linear subspace of $\mathfrak{H}$ consisting of all λ such that $w\lambda = -\lambda$. We see that

$$\mathfrak{H} = \bigcup_{w \in W} \mathfrak{H}_w,$$

and, since W is finite, it follows that $\mathfrak{H} = \mathfrak{H}_{w^0}$ for some $w^0 \in W$. It is clear that $w^0 = -\mathrm{Id}$.

3.6. By analyzing the root systems of symmetric spaces, it is easy to give a complete list of all symmetric spaces of noncompact type that satisfy the condition $-\mathrm{Id} \in W$.

Since the Weyl group W is generated by the reflections

$$s_\alpha = \lambda - 2\frac{\langle \lambda, \alpha\rangle}{\langle \alpha, \alpha\rangle}\alpha$$

of the Cartan subalgebra corresponding to the hyperplanes orthogonal to α, it depends on the type of the root system whether a reflection belongs to the Weyl group or not.

Let us consider types of root systems of semisimple Lie algebras one by one (see, for example, [20]).

(1) Type A_n. The Cartan subalgebra is represented by the hyperplane $x_1 + \cdots + x_{n+1} = 0$ in $\mathbb{R}^n$; the root system is

$$\alpha_{ik} = e_i - e_k,$$

where $e_1, \ldots, e_{n+1}$ is a standard basis in $\mathbb{R}^{n+1}$. The reflection $s_{\alpha_{ik}}$ permutes coordinates i and k. If $n > 1$, we have $(-\mathrm{Id}) \notin W$.

(2) Types B_n, C_n. The root systems for these two types include the vectors of the orthogonal basis e_i, $i = 1, \ldots, n$, so $(-\mathrm{Id}) \in W$.

(3) Type D_n. The root system consists of the vectors

$$\alpha_{ik}^{\pm,\pm} = \pm e_i \pm e_k.$$

If n is even, $n = 2m$, the root system

$$\alpha_{2i-1,\,2i}^{++}, \; \alpha_{2i-1,\,2i}^{+-}, \qquad i = 1, \ldots, m,$$

is an orthogonal basis in $\mathbb{R}^n$; therefore, $(-\mathrm{Id}) \in W$.

The reflections $s_{\alpha_{ik}^{+-}}$, $s_{\alpha_{ik}^{-+}}$ interchange coordinates i and k; the reflections $s_{\alpha_{ik}^{++}}$, $s_{\alpha_{ik}^{--}}$ also change their signs. If n is odd, it is impossible to convert, say, $(1, \ldots, 1)$, into the opposite vector by a composition of such transformations, because each transformation changes the signs of two components. Thus, if n is odd, $-\mathrm{Id} \notin W$.

(4) The mixed type root system BC_n contains the basis $\{e_i\}$; therefore, $(-\mathrm{Id}) \in W$.

We now consider exceptional root systems.

(5) E_6. The Cartan subalgebra is now the hypersurface $x_1 + \cdots + x_6 = 0$ in $\mathbb{R}^7$. The roots are

$$\alpha_{ik} = e_i - e_k, \qquad i, k = 1, \ldots, 6,$$

and

$$\beta_1 = \sqrt{2}e_7, \qquad \beta_2 = \tfrac{1}{2}(\pm e_1 \pm \cdots \pm e_6).$$

The reflections $s_{\alpha_{ik}}$ interchange two of the first six coordinates, and s_{β_1} changes the sign of the seventh coordinate. No sequence of such operations together with s_{β_2} can change signs of all seven coordinates, so $-\mathrm{Id} \notin W$ (see N. Bourbaki, "Groupes et Algèbres de Lie", p. 271 of the Russian translation).

(6) E_7. The Cartan subalgebra is represented by the hypersurface $x_1 + \cdots + x_8 = 0$ in $\mathbb{R}^8$. Among the roots there are seven linearly independent pairwise orthogonal vectors $e_1 - e_2$, $e_3 - e_4$, $e_5 - e_6$, $e_7 - e_8$, $\frac{1}{2}(e_1 + e_2 - e_3 - e_4 + e_5 + e_6 - e_7 - e_8)$, $\frac{1}{2}(e_1 + e_2 + e_3 + e_4 - e_5 - e_6 - e_7 - e_8)$, $\frac{1}{2}(e_1 + e_2 - e_3 - e_4 - e_5 - e_6 + e_7 + e_8)$. Therefore, $-\mathrm{Id} \in W$.

(7) E_8. In this case one has also $-\mathrm{Id} \in W$, since the root system contains the orthogonal basis obtained by adding the vector $e_1 + \cdots + e_8$ to the set of vectors corresponding to E_7.

(8) F_4. The basis vectors e_1, e_2, e_3, e_4 are roots; therefore, $-\mathrm{Id} \in W$.

(9) G_2. The vectors e_1, $\sqrt{3}e_2$ are roots; therefore, $-\mathrm{Id} \in W$.

Thus the condition $-\mathrm{Id} \notin W$ does not hold for symmetric spaces with root systems of types A_n, D_{2n-1} $(n > 1)$, E_6. Here is a list of irreducible symmetric spaces of noncompact type with these root systems (see [28]):

$$\begin{aligned}
&SL(n+1, \mathbb{C})/SU(n+1), \qquad n > 1 \quad (A_n);\\
&SO(4n+2, \mathbb{C})/SO(4n+2) \qquad (D_{2n+1});\\
&E_6^{\mathbb{C}}/E_6 \qquad (E_6);\\
&SL(n, \mathbb{R})/SO(n), \qquad n > 1 \quad (A_n);\\
&SU^*(2n)/Sp(n), \qquad n > 2 \quad (A_{n-1});\\
&SO_0(2n+1, 2n+1)/(SO(2n+1) \times SO(2n+1)) \qquad (D_{2n+1});\\
&E\,\mathrm{I} \qquad (E_6);\\
&E\,\mathrm{IV} \qquad (A_2).
\end{aligned} \tag{$*$}$$

The type of root system corresponding to the space is enclosed in parentheses. All the Hermitian symmetric spaces satisfy the condition $-\mathrm{Id} \in W$:

$$\begin{aligned}
&SU(p, q)/S(U_p \times U_q) \qquad (BC_q \text{ for } p \neq q \text{ and } C_q \text{ for } p = q);\\
&SO_0^*(2n)/U(n) \qquad (C_{n/2} \text{ for even } n,\ BC_{[n/2]} \text{ for odd } n);\\
&SO_0(p, 2)/SO(p) \times SO(2) \qquad (B_2);\\
&Sp(n, \mathbb{R})/U(n) \qquad (C_n);\\
&E\,\mathrm{III} \qquad (BC_2);\\
&E\,\mathrm{VII} \qquad (C_3).
\end{aligned}$$

This is also easy to derive from the general properties of these spaces, without using the classification of symmetric spaces.

We summarize our conclusions in the following theorem.

3.7. THEOREM. *Let X be a connected irreducible symmetric space of noncompact type and $I_0(X)$ a connected component of the isometry group of X. Then the following conditions are equivalent:*

(1) *Every closed $I_0(X)$-invariant subspace of $L^2(X)$ is closed under complex conjugation.*

(2) *All positive definite spherical functions on X are real valued.*

(3) *The Weyl group W contains the reflection $-\mathrm{Id}$.*

(4) *X has a root system other than A_n, D_{2n-1}, $n > 1$; E_6.*

(5) *X is not locally isometric to any of the symmetric spaces listed in $(*)$.*

§4. Invariant algebras on symmetric spaces of noncompact type

In this section we use the results of §3 to describe isometry-invariant function algebras on symmetric spaces of noncompact type.

4.1. LEMMA. *Let X be an irreducible connected symmetric space of noncompact type and $\mathscr{F}$ an I_0-invariant set of continuous functions on X. If $\mathscr{F}$ contains nonconstant functions, it separates the points of X.*

PROOF. Our argument resembles that of [104]. The main idea is to apply a theorem of Kobayashi [78], according to which the set of fixed points of the isotopy group of a Riemannian space of negative curvature is connected.

Fix a base point $o \in X$, and let K be the stationary subgroup of $G = I_0(X)$ at o. Then K is a maximal compact subgroup of G. Set

$$X_a = \{x \in X : f(x) = f(a), \ f \in \mathscr{F}\},$$

and let H denote the subgroup of G consisting of all $g \in G$ such that $gX_0 \subset X_0$.

Since X_0 is closed, H is closed in G and, hence, is a Lie group. It is clear that $K \subset H$. For the Lie algebras,

$$\mathfrak{K} \subset \mathfrak{H} \subset \mathfrak{G}.$$

As X is irreducible, $\mathfrak{K}$ is a maximal subalgebra of $\mathfrak{G}$ ([28], p. 337 of Russian translation). Consequently $\mathfrak{H}$ is either $\mathfrak{K}$ or $\mathfrak{G}$. In the latter case $H = G$, so $X_0 = X$ and $\mathscr{F} = \mathbb{C}$. If $\mathfrak{K} = \mathfrak{H}$, then K and H are locally homomorphic, and therefore the fiber $X_0 = H/K$ is discrete. Let a be a fixed point of K and $s \in X_a$. Consider the (continuous) map: $K \ni k \to ks \in X$. Since $\mathscr{F}$ is invariant, the group G preserves the fibering $\{X_a\}$, $a \in A$, and thus $ks \in X_a$, that is, the range of the map is contained in X_a. Since K is connected and X_a is discrete, it follows that the map is the identity, i.e., each point of X_a is a fixed point of K.

Thus if we let Z denote the set of fixed points of K, then Z is a union of fibers

$$Z = \bigcup_{a \in Z} X_a.$$

Let H_Z denote the group consisting of all $g \in G$ such that $gZ \subset Z$. It is clear that H_Z acts transitively on Z, because fibers are mapped to fibers and G is transitive on X. Obviously $K \subset H_Z$. By the irreducibility of the symmetric space X, we again conclude that K and H_Z are locally homomorphic (equality $H_Z = G$ is impossible, because it would imply $K = \{e\}$); therefore, $Z = H_Z/K$ is discrete. But by Kobayashi's theorem Z is connected, so $Z = \{e\}$. Thus $X_0 = \{0\}$, i.e., $\mathscr{F}$ separates the points of X.

4.2. THEOREM. *Let X be a connected irreducible symmetric space of noncompact type. Suppose that X satisfies one of the equivalent conditions* (3)–(5)

of Theorem 3.5. *Further, let* A *be an* $I_0(X)$*-invariant subalgebra of* $C_0(X)$ *such that, for some natural number* p,

$$A \cap L^p(X) \neq \{0\}.$$

Then $A = C_0(X)$.

PROOF. Let $f \in A \cap L^p(X)$, $f \not\equiv 0$. Then $f^p \in L^1(X)$ and, since f is bounded, it follows that $f^p \in L^2(X)$. Thus $A \cap L^2(X) \neq \{0\}$. Set $Y = \mathrm{cl}_{L^2}[A \cap L^2(X)]$, and let $A_0 = \mathrm{cl}_{C_0(X)}[C_0(X) \cap Y]$. Then A_0 is an $I_0(X)$-invariant closed subalgebra of $C_0(X)$. By Theorem 3.5, Y is closed under complex conjugation, hence so is A_0. By Lemma 4.1, A_0 separates points on X. Now, by the Stone-Weierstrass theorem, $A_0 = C_0(X)$. Lemma 3.1 implies $A = C_0(X)$.

COROLLARY. *Let* X *be as in Theorem* 4.2. *If* $h \in C_0(X)$ *is a nonzero function on* X *some degree of which is absolutely integrable, then any function in* $C_0(X)$ *can be uniformly approximated by polynomials in translations* $h \circ \psi$, $\psi \in I_0(X)$.

CHAPTER II

Translation Invariant Function Spaces and Function Algebras on Noncompact Lie Groups

The main subject studied in this chapter is the translation-invariant function algebras on noncompact semisimple Lie groups and the Heisenberg group.

In §1 we will prove an irreducibility theorem for semisimple Lie groups for two-sided invariant algebras satisfying certain natural regularity conditions. In particular, these regularity conditions will allow us to eliminate a certain restriction on the underlying symmetric space (that the Weyl group contains a symmetry) in Theorem 4.2.

In §2, we will study invariant function algebras and function spaces on the Heisenberg group—a nilpotent Lie group of degree 2. Contrary to the situation in semisimple groups, in this case the class of invariant algebras is quite rich.

§1. Invariant algebras on noncompact semisimple Lie groups

Throughout this section G will be a connected noncompact semisimple Lie group with finite center, K a maximal compact subgroup, and $X = G/K$ the symmetric space of left cosets. We denote by π the natural projection of G to X and let $o = \pi(e)$.

This section contains proofs of the following two theorems.

1.1. THEOREM. *Let A be a closed G-invariant subalgebra of $C_0(X)$. Suppose that*:

(1) *$A \cap L^1(X)$ is uniformly dense in A*;

(2) *the linear span of the translates of K-invariant functions lying in $A \cap L^1(X)$ is dense in A in the norm of $L^1(X)$. Then either $A = \{0\}$ or $A = C_0(X)$.*

REMARK. Condition (1) was also adopted in Theorem 1.4.2. As for condition (2), it follows from Proposition 3.3, Chapter I, that the linear combinations of translates of K-invariant functions are dense in the $L^2(X)$-norm. Here, however, we have assumed the validity of this approximation property in the L^1-norm.

1.2. THEOREM. *Let A be a closed subalgebra of $C_0(G)$ invariant under both left and right translations. Suppose that $A \cap L^1(G)$ is uniformly dense in A and that the cyclic linear span of the K-bi-invariant functions that belong to $A \cap L^1(G)$ is dense in $A \cap L^1(G)$ in the L^1-norm. Then either $A = \{0\}$ or $A = C_0(G/Z)$, where Z is a subgroup of the center of G.*

We say that function a f is K-bi-invariant if $f(k_1 x k_2) = f(x)$, $k_1, k_2 \in K$.

COROLLARY. *Let G be a connected semisimple real Lie group with trivial center and f a nonzero K-bi-invariant function belonging to $L^1(G)$. Then every element of $C_0(G)$ can be uniformly approximated by polynomials in translates of f.*

Theorem 1.2 follows from Theorem 1.1. The plan of the proof of Theorem 1.1 is as follows. First we use the Iwasawa decomposition $G = ZHK$ to represent G as a product of a nilpotent group Z, an (abelian) Cartan subgroup H, and a maximal compact subgroup K, together with the Harish-Chandra transformation, based on averaging over Z, to associate to each algebra A a translation-invariant subspace A^H of functions on H. We single out a subspace A^H of A that is mapped by the Harish-Chandra transformation into the space of integrable functions on H. Then we show that A^H has integrable multipliers; therefore, the Beurling spectrum of A^H is invariant under addition of some open subsemigroup of the dual group $\widehat{H}$. Since the spectrum is invariant under the action of the Weyl group, it must be the whole of $\widehat{H}$. Harish-Chandra's formula relating the K-spherical Fourier transform on X to the classical Fourier transform on the Cartan subalgebra implies that the initial space A coincides with $C_0(X)$.

1.3. Thus, consider the Iwasawa decomposition $G = ZHK$. Let $\gamma(x)$ denote the (uniquely defined) element k in the representation $x = zho$, $z \in Z$, $h \in H$. Let $\gamma^* : f \to f \circ \gamma$ be the adjoint map of function spaces. Also set $\rho(h) = \det \operatorname{Ad}_S(h^{-1})$, where Ad_S is the adjoint representation of the group $S = ZH$.

Now consider the averaging operator corresponding to the subgroup Z:

$$(S_Z f)(h) = \int_Z f(zho)\, dz,$$

and let V denote the Harish-Chandra transformation, $V = \rho^{1/2} S$. Using the formula for G-invariant measure on X [28], we have

$$\int_X f(x)\, dx = \int_{Z \times H} f(zho)\rho(h)\, dh\, dz = \int_H (S_Z f)(h)\rho\, dh. \tag{1}$$

Hence S_Z maps $L^1(X)$ into $L^1(H, \rho dh)$ and V maps $L^1(X)$ into $L^1(H, \rho^{1/2} dh)$.

1.4. Lemma. *The Harish-Chandra transformation maps the G-invariant linear span of $L^1_K(X)$ into $L^1(H)$.*

Proof. By the Cauchy-Bunykovskiĭ inequality,

$$L^1(H, \rho^{(-1/2)}dh) \cap L^1(H, \rho^{1/2}dh) \subset L^1(H).$$

Thus it is enough to show that $Vf \in L^1(H, \rho^{-1/2}dh)$, or, in other words, that $S_Z f \in L^1(H)$ for any function f from this linear span of $L^1_K(X)$.

First suppose $f \in L^1_K(X)$. Lift f to G and invert the argument: $f^*(g) = (f \circ \pi)(g^{-1})$. The function $f \circ \pi$ is constant on the two-sided cosets of K; therefore, so is f^*. Since K is compact and G is unimodular, it follows that $f^* \in L^1(G)$.

Since H is contained in the normalizer of Z, we can write

$$\int_{Z\times H} |f(zho)|\,dz\,dh = \int_{H\times Z} |f^*(hz)|\,dz\,dh = \int_{Z\times H} |f^*(hzh^{-1}h)|\,dh\,dz$$
$$= \int_{Z\times H} |f^*(zh)|\rho(h)\,dz\,dh = \int_G |f^*(g)|\,dg.$$

Since $\int_G |f^*(g)|\,dg < \infty$, we have $(S_Z f)(h) = \int_Z f(zh)\,dz \in L^1(H)$.

If f is a translate of a K-invariant function, then the inclusion $S_Z f \in L^1(H)$ follows from the following relation between translation operators and S_Z, which can easily be checked:

$$S_Z T_g = \rho(h^{-1}) T_h S_Z T_k, \tag{2}$$

where $g = zhk$ is the Iwasawa decomposition of the element $g \in G$. There is also an equivalent formula

$$VT_g = \rho^{(-1/2)}(h) T_h V T_k. \tag{2'}$$

Now let A be an algebra that satisfies all the conditions of Theorem 2.1.2, $A \neq \{0\}$.

1.5. Lemma. *Let $f \in A \cap L^1(X)$, and suppose that $S_Z(|f|)$ is bounded on H. Then the product*

$$(\gamma^* S_Z f) \cdot g$$

is an element of A for any $g \in A$; in other words, $\gamma^* S_Z f$ is a multiplier of A.

Proof. It is enough to show that

$$\int_X (\gamma^* S_Z f) \cdot g\,d\mu = 0$$

for any measure $\mu \in A^{\perp}$ on X belonging to the annihilator $A^{\perp}$ of A in the dual space of $C_0(X)$. To that end, consider a compact covering $\{Z_m\}$ of the group Z, and set

$$f_m = \int_{Z_m} (T_z f)\, dz.$$

The estimate for the uniform norms

$$\|f_m g\|_\infty \le \|\gamma^* S_Z(|f|)\|_\infty \cdot \|g\|_\infty$$

and the assumption that the first factor is bounded imply that the sequence $f_m g$ is uniformly bounded. In addition, $f_m g \in A$ for every m, and the sequence is pointwise convergent to $(\gamma^* S_Z f)\cdot g$. Using Lebesgue's dominated convergence theorem and letting $m \longrightarrow \infty$ in the equality

$$\int_X f_m g\, d\mu = 0,$$

we obtain the assertion.

1.6. Define the subset

$$A^0 = \{f \in A \cap L^1(X) :\ S_Z(|f|) \in L^1(H)\}.$$

By Lemma 1.4, A^0 contains $A \cap L^1_K(X)$. It follows from (1) that A^0 is G-invariant. It is of course a linear subspace and an ideal in A, because of the inequality

$$S_Z(|fg|) \le \|g\|_\infty \cdot S_Z(|f|).$$

Let $f \in A^0$, then $Vf \in L^1(H,\ \rho^{1/2} dh)$. In addition, by the definition of A^0,

$$Vf = \rho^{1/2} S_Z f \in L^1(H,\ \rho^{-1/2} dh).$$

The last two inclusions imply that $Vf \in L^1(H)$, so $V(A^0) \subset L^1(H)$. Note also that, by (1), the space $V(A^0)$ is invariant under translation by H.

LEMMA. *Let A^H denote the uniform closure of the set $C_0(H) \cap V(A^0)$ and $M(A^H)$ the algebra of multipliers of A^H in $C_0(H)$. Then*

$$M(A^H) \cap L^1(H) \ne \{0\}.$$

PROOF. There exists a function $f_0 \in A^0$ such that $S_Z f_0 \not\equiv 0$. Indeed, as shown in Lemma 1.4, $A \cap L^1_K(X) \subset A^0$. If $S_Z(A^0) = \{0\}$, then $V(A \cap L^1_K(X)) = \{0\}$, but then the Harish-Chandra formula (see below) implies that $A \cap L^1_K(X) = \{0\}$, and in that case $A = \{0\}$.

Choose a continuous compactly supported function χ on H such that

$$(\rho^{-1}\chi) * S_Z f_0 \not\equiv 0.$$

Formula (2) implies

$$S_Z f = \rho^{-1}\chi * S_Z f_0, \tag{3}$$

where

$$f(x) = (\chi *_H f_0)(x) = \int_H \chi(h) f(h^{-1}x)\, dh,$$

so that $S_Z f$ belongs to $L^1(H) \cap C_0(H)$ as the convolution of a compactly supported function by an integrable function. In addition, the function $S_Z(|f|)$ is bounded, because

$$\|S_Z(|f|)\|_\infty \le \|\rho^{-1}\chi\|_\infty \cdot \|S_Z(|f|)\|_{L^1(H)} < \infty.$$

It follows from Lemma 1.5 that $\gamma^* S_Z f$ is a multiplier of A. Then $S_Z f$ is a multiplier of A^H. Indeed, if $g \in V(A^0)$, in other words, g can be represented as $g = V g_0$ for $g_0 \in A^0$, then $(\gamma^* S_Z f) \cdot g \in A^0$. Since $\gamma^* S_Z f$ is constant on the orbits of Z, it follows that $(S_Z f) \cdot g = V\left[(\gamma^* S_Z f) \cdot g_0\right]$, and therefore $(S_Z f) \cdot g \in V(A^0)$.

COROLLARY. *Let A^H be a translation-invariant space and $\sigma(A^H)$ denote its Beurling spectrum (i.e., the set of group characters that belong to the weak closure of A^H). Then there exists an open semigroup $\mathscr{P}$ of the dual group $\widehat{H}$ such that*

$$\mathscr{P} + \sigma(A^H) \subset \sigma(A^H).$$

PROOF. Set $\mathscr{P} = \operatorname{int} \sigma(M(A^H))$. Since $M(A^H)$ is a translation-invariant algebra, it follows that $\mathscr{P}$ is a semigroup (see, e.g., [81]). By Lemma 1.6, $M(A^H) \cap L^1(H) \neq \{0\}$; hence, $\mathscr{P} \neq \varnothing$. The desired inclusion now follows from the fact that the Beurling spectrum of the product of two bounded functions is contained in the closure of the sum of their spectra [81].

1.7. LEMMA. *Let A_K^H denote the uniform closure of the space $C_0(H) \cap \Phi$, where $\Phi = [V(A \cap L_K^1(X))]_H$ is the linear span of the set of translates of elements of $V(A \cap L_K^1(X))$. Then the Beurling spectra $\sigma(A_K^H)$ and $\sigma(A^H)$ coincide.*

PROOF. It follows directly from $(2')$ that A_K^H is the uniform closure of the set

$$C_0(H) \cap V\left([A \cap L_K^1(X)]_G\right).$$

By Lemma 1.4, the latter is contained in A^H; hence, $A_K^H \subset A^H$ and

$$\sigma(A_K^H) \subset \sigma(A^H).$$

Suppose the spectra do not coincide, and let $\lambda \in \sigma(A^H) \backslash \sigma(A_K^H)$. There exists a neighborhood U of λ whose closure does not intersect $\sigma(A_K^H)$. Since the spectrum of a convolution is contained in the intersection of the spectra of the factors, we may use convolution to construct a function $g_0 \in C_0(H) \cap V(A^0)$ whose Beurling spectrum does not intersect $\sigma(A_K^H)$. Then $g_0 * g = 0$ for any $g \in A_K^H$. We have $g_0 = V f_0$ for some $f_0 \in A^0$. Thus

$$(V f_0) * g = 0, \qquad g \in A_K^H \tag{4}$$

(the convolution is calculated on H). Take a function $\psi \in A \cap L_K^1(X)$. By the corollary to Chapter I, Lemma 3.2, the involution

$$\psi^{\#}(go) = \overline{\psi(g^{-1}o)}$$

belongs to $A \cap L_K^1(X)$. Set $g = V\psi^{\#}$ in (4), and let the argument of the convolution on the left of (4) be equal to the unit element e of H. Then, by (2′),

$$\begin{aligned}\int_H (Vf_0)(h)\overline{(V\psi)(h)}\,dh &= \int_H (Vf_0)(h)(VT_h\psi^{\#})(e)\rho^{1/2}(h)\,dh \\ &= \int_H (Vf_0)(h)(V\psi^{\#})(h^{-1})\,dh = 0.\end{aligned} \tag{5}$$

One can replace f_0 in (5) by any translate $T_h f_0$, $h \in H$. Therefore, applying the translation to the second factor instead of the first, and using (2′), we see that all the integrals in (5) will still be zero if ψ is replaced by $T_g\psi$, where g is an arbitrary element of G. Using this remark, and using formula (1) to transform the expression on the left of (5) into an integral over X, we obtain

$$\int_X (\gamma^* S_Z f_0)\overline{T_g\psi}\,dx = 0. \tag{6}$$

We could have started with a convolution $\chi * f_0$ instead of f_0, where χ is a continuous function with compact support. Therefore, using (2), we may assume that $\gamma^* S_Z f_0$ is bounded on X. Using approximations in L^1-norm of f_0 by linear combinations of $T_g\psi$ which exist by condition (2), we see that

$$\int_X (\gamma^* S_Z f_0)\overline{f_0}\,dx = 0.$$

But now the formula for a G-invariant measure implies

$$\begin{aligned}\int_H |(S_Z f_0)(h)|^2 \rho(h)\,dh &= \int_X S_Z\left[(\gamma^* S_Z f_0)\cdot\overline{f_0}\right]\rho(h)\,dh \\ &= \int_X (\gamma^* S_Z f_0)\overline{f_0}\,dx = 0.\end{aligned}$$

Since $\rho > 0$, we have $S_Z f_0 = 0$. But this contradicts the choice of f_0. Thus our assumption $\sigma(A_K^H) \subsetneq \sigma(A^H)$ is false, and the lemma follows.

1.8. Let $\mathfrak{H}$ denote the Cartan subalgebra, i.e., the Lie algebra of H. The following formula, due to Harish-Chandra, relates the Fourier transforms on a symmetric space X to Fourier transforms on a locally compact abelian group [28]:

$$\hat{f}(\nu) = \widehat{Vf}(\exp \nu), \qquad \nu \in \mathfrak{H}.$$

LEMMA. *Let* $\widehat{\exp} : \widehat{H} \to \widehat{\mathfrak{H}} \cong \mathfrak{H}$ *be the character homomorphism dual to the exponential map* $\exp : \mathfrak{H} \to H$. *Then*

$$\widehat{\exp}\, \sigma(A_K^H) = \mathfrak{H}.$$

PROOF. Let $\lambda_1, \dots, \lambda_n \in \mathfrak{H}$ be a system of linearly independent roots of the Lie algebra of G with respect to $\mathfrak{H}$. The Weyl group is generated by the reflections of $\mathfrak{H}$ in the hyperplanes

$$\Pi_i = \{\langle \lambda, \lambda_i \rangle = 0\}.$$

The set $\widehat{\exp}\, \sigma(A_K^H)$ is invariant under these reflections. On the other hand, by the corollary to Lemma 1.6, it is invariant under the addition of some open semigroup $\mathscr{P}$ of $\mathfrak{H}$. Any open semigroup of the finite-dimensional space $\mathfrak{H}$ contains a set $a + \mathfrak{J}$, where $a \in \mathfrak{H}$ and $\mathfrak{J}$ is an open cone in $\mathfrak{H}$. Since the vectors $\lambda_1, \dots, \lambda_n$ form a basis of $\mathfrak{H}$, we have

$$\sum_{w \in W} w(\mathfrak{J}) = \mathfrak{H}.$$

The lemma now follows from the inclusions

$$\mathfrak{H} = \sum_{w \in W} w(a + \mathfrak{J}) + \widehat{\exp}\, \sigma(A_K^H) \subset \sum_{w \in W} w(\mathscr{P}) + \widehat{\exp}\, \sigma(A_K^H)$$
$$\subset \widehat{\exp}\, \sigma(A_K^H) \subset \mathfrak{H}.$$

PROOF OF THEOREM 1.1. Let E denote the closure $A \cap L_K^2(X)$ in $L^2(X)$. By Lemma 3.2, Chapter I, $E = L_{K,\xi}^2(X)$ for some measurable set $\xi \subset \mathfrak{H}/W$. The space A_K^H is the uniform closure of the set $\Phi = [V(A \cap L_K^1(X))]_H$ which consists of functions integrable on H. Therefore, $\sigma(A_K^H)$ is the closure of the union η of supports of the Fourier transforms of the functions Vf, $f \in A \cap L_K^1(X)$. By Lemma 1.8, $\widehat{\exp}\, \eta$ is dense in $\mathfrak{H}$. On the other hand, Harish-Chandra's formula implies that the quotient space $\widehat{\exp}\, \eta / W$ is contained in ξ; therefore, ξ is dense in $\mathfrak{H}/W$.

Consider the annihilator $A^\perp$ of A in the measure space dual to $C_0(X)$. Let $\mu \in A^\perp$ be a measure of the form given by

$$\mu = \varphi\, dx, \qquad \varphi \in L^1(X) \cap L^\infty(X) \cap C_0(X).$$

For any element $g \in G$, the average $S_K T_g \varphi$ of φ over the group K is orthogonal to A:

$$\int_X f(x)(S_K T_g \varphi)(x)\, dx = 0, \qquad f \in A \cap L_K^2(X).$$

Since $S_K T_g \varphi \in L^2(X)$ and $A \cap L^2_K(X)$ is L^2-dense in $E = L^2_{K,\xi}$, the equality also holds for $f \in E$. By the Plancherel formula,

$$\int_{\mathfrak{H}/W} \hat{f}(\lambda)\, \overline{\widehat{S_K T_g \overline{\varphi}}(\lambda)}\, d\lambda = 0, \qquad f \in L^2_{K,\xi}(X).$$

But ξ is dense in $\mathfrak{H}/W$, and $\widehat{S_K T_g \overline{\varphi}}$ is continuous on $\mathfrak{H}$, as the Fourier transform of an integrable function. Thus $\widehat{S_K T_g \overline{\varphi}} = 0$. Therefore $S_K T_g \overline{\varphi} = 0$, in particular,

$$\overline{\varphi}(g^{-1} o) = (S_K T_g \overline{\varphi})(g^{-1} o) = 0.$$

Since g is an arbitrary element of G, it follows that $\varphi = 0$. Using convolutions of measures with δ-sequences of continuous functions with compact support on X, one easily shows that the measures of the form φdx, where $\varphi \in L^1(X) \cap L^\infty(X) \cap C_0(X)$, which annihilate A form a weakly dense subset of $A^\perp$ (cf. Lemma 1.3.1). Hence $A^\perp = \{0\}$ and $A = C_0(X)$.

Proof of Theorem 1.2. Let A^K denote the subalgebra of A consisting of right K-invariant functions, i.e., functions f such that $f(xk) = f(x)$, $k \in K$. We suppose that $A \neq \{0\}$. Then the assumptions of the theorem imply $A^K \neq \{0\}$. It is also clear that A^K, viewed as a function algebra on the quotient space $X = G/K$, satisfies all the assumptions of Theorem 1.1. Hence $A^K = C_0(X)$.

Suppose first that A separates points on G. Let $A_1 = A \oplus \mathbb{C}$ be the linear span of A and constant functions. Consider the maximal antisymmetry set (see §2.1 in Chapter I) K_e of A_1 that contains the unit element of G. We do not include in this set the point at infinity of the one-point compactification G_∞ of G (of course, A_1 is defined on G_∞). The partition of G into maximal antisymmetry sets is preserved under left and right translation, since the algebra is invariant. Let $a \in K_e$; then aK_e and K_e are maximal antisymmetry sets with a common element a, so $aK_e = K_e$. Hence, K_e is a group. Since K_e is mapped by inner automorphisms into itself, it is a normal subgroup. It clearly follows from the equality $A^K = C_0(G/K)$ that $K_e \subset K$. The Lie algebra of K does not contain nonzero ideals of the Lie algebra of G [28, Theorem 3.3]; therefore, the Lie algebra of K_e is trivial and K_e is a discrete normal group. Since G is connected, it follows that K_e is contained in the center $Z(G)$ of G and K_e is finite. Since A separates points, K_e must be the singleton $\{e\}$, otherwise it is easy to construct a function in A which is real and nonconstant on K_e.

Thus, all maximal antisymmetry sets are singletons, since any such set is a translate of K_e. This implies that A is closed under complex conjugation, and by the Stone-Weierstrass theorem $A = C_0(G)$.

If A does not separate points on G, we define

$$Z = \{g \in G:\ f(g) = f(e),\ f \in A\}.$$

The equality $A^K = C_0(G/K)$ implies $Z \subset K$. Since A is invariant, Z is a (discrete) normal subgroup of G, so $Z \subset Z(G)$. The algebra A, viewed as an algebra on the quotient group G/Z, separates points, and we have already shown that then $A = C_0(G/Z)$.

§2. Affine-invariant function spaces and function algebras on the Heisenberg group

We showed in the preceding section that in the case of semisimple Lie groups invariant algebras are in fact nothing but the algebra of all continuous functions. However, for groups with nontrivial center the situation is different.

In this section we study invariant function spaces and function algebras on the Heisenberg group H^n, which is one of the most interesting nilpotent Lie groups from the point of view of analysis. The resulting classification of these spaces and algebras turns out to be quite rich. Here one obtains spaces of boundary values of analytic functions (for an appropriate realization of the Heisenberg group), which are of special importance for the classification of invariant spaces on H^n.

Let us describe the contents of the section in more detail. We will be concerned with the multidimensional generalization of the following result of de Leeuw and Mirkil [81]. Let $A(\mathbb{R})$ denote the Phragmén-Lindelöf algebra, that is, the set of boundary values on the real axis of functions that are analytic in the upper half-plane and vanish at infinity. Then $A(\mathbb{R})$ and its complex conjugate $\overline{A}(\mathbb{R})$ are the only closed nonzero proper subalgebras of $C_0(\mathbb{R})$ that are invariant under translations and contain a dense subset of integrable functions and an approximate unit. This follows from the description [81] of translation-invariant function algebras on locally compact abelian groups, in particular, on $\mathbb{R}^n$, in terms of Beurling spectra—subsemigroups of the dual group.

Replacing $\mathbb{R}$ with $\mathbb{R}^n$ in the bounded represention of the upper half-plane by a complex disk corresponds to replacing the disk with a polydisk. We are interested in generalizations of the complex ball type, which corresponds to replacing $\mathbb{R}$ by the Heisenberg group H^n.

The group H^n may be defined as the set $\mathbb{R} \times \mathbb{C}^{n-1}$ endowed with the group operation

$$(x', \zeta') \cdot (x, \zeta) = (x + x' - 2\,\mathrm{Im}\langle \zeta, \zeta' \rangle, \zeta + \zeta'),$$

where $\langle \zeta, \zeta' \rangle = \sum_{k=1}^{n-1} \zeta_k \overline{\zeta}'_k$. The analogue of the upper half-plane is then the Siegel domain

$$\Omega^n = \{(z, \zeta) \in \mathbb{C} \times \mathbb{C}^{n-1} : \mathrm{Im}\, z > |\zeta|^2\},$$

where $|\zeta|^2 = \langle \zeta, \zeta \rangle$.

The map $\gamma : (x, \zeta) \to (x + i|\zeta|^2, \zeta)$ determines a one-to-one correspondence between H^n and the boundary

$$\partial\Omega^n = \{(z, \zeta) \in \mathbb{C} \times \mathbb{C}^{n-1} : \operatorname{Im} z = |\zeta|^2\}$$

of Ω^n. Thus $\partial\Omega^n$ is endowed with a group structure, with the multiplication

$$(z', \zeta') \cdot (z, \zeta) = (z + z' + 2i\langle\zeta, \zeta'\rangle, \zeta + \zeta'),$$

so that γ is a group isomorphism. The left translations

$$\omega_{(z', \zeta')}(z, \zeta) = (z', \zeta')^{-1} \cdot (z, \zeta), \qquad (z', \zeta') \in \partial\Omega^n,$$

constitute a group of analytic automorphisms of Ω^n isomorphic to H^n.

The Cayley transformation

$$\Phi(w) = \left(i\frac{1 + w_1}{1 - w_1}, i\frac{w_2}{1 - w_1}, \dots, i\frac{w_n}{1 - w_1}\right)$$

is an analytic map of the unit ball B^n of $\mathbb{C}^n$ onto the domain Ω^n. Therefore, H^n is isomorphic to a subgroup of the group $\operatorname{Aut}(B^n)$ of analytic automorphisms of the ball.

In this section we will describe subalgebras of $C_0(H^n)$ that are invariant under transformations, including translations and affine automorphisms.

Let us outline the results. We will define two algebras on the boundary $\partial\Omega^n$ of the Siegel domain Ω^n that are similar to Phragmén-Lindelöf algebra on the real axis. Let $A(\partial\Omega^n)$ denote the algebra of boundary values on $\partial\Omega^n$ of functions which are holomorphic in Ω^n, continuous in $\Omega^n \cup \partial\Omega^n \cup \{\infty\}$, and vanishing at infinity. The algebra $B(\partial\Omega^n)$ is defined in the same way but starting from functions which are holomorphic in z instead of holomorphic functions in (z, ζ).

Using γ to push the functions forward from $\partial\Omega^n$ to H^n, we obtain the algebras $A(H^n)$, $B(H^n)$ on H^n. These algebras are invariant under left translations, and $B(H^n)$ is invariant also under right translations. In addition, both algebras are invariant under unitary transformations in the space of variables $\zeta_1, \dots, \zeta_{n-1}$ and under anisotropic dilatations $\delta_t(x, \zeta) = (t^2 x, t\zeta)$, $t > 0$.

Our main result is a classification of translation-invariant, polydisk (i.e., invariant under the action of the torus group $\zeta_k \to e^{i\varphi_k}\zeta_k$, $k = 1, \dots, n-1$), closed subalgebras of $C_0(H^n)$ that contain an approximate unit and a dense subset of integrable functions. All such algebras, except $\{0\}$ and $C_0(H^n)$, lie between $A(H^n)$ and $B(H^n)$ or between $\overline{A}(H^n)$ and $\overline{B}(H^n)$. They are completely described in terms of harmonic analysis on the Heisenberg group. The class of invariant subalgebras between $A(H^n)$ and $B(H^n)$ corresponds to a certain class of subsemigroups of $\mathbb{R}_+ \times Z_+^{n-1}$, where $\mathbb{R}_+ = [0, \infty)$ and $Z_+ = \{0, 1, \dots\}$. The corresponding complex conjugate algebras lie between $\overline{A}(H^n)$ and $\overline{B}(H^n)$.

The existence of these two discrete series of algebras is specific to the multidimensional case. When $n = 1$ the group degenerates into $\mathbb{R}$, the algebras

$A(H^n)$ and $B(H^n)$ coincide, and we obtain only four invariant algebras, in agreement with the theorem of de Leeuw and Mirkil. In the multidimensional case the extra condition of invariance with respect to the anisotropic dilatations δ_t excludes infinite series. It turns out that there are six anisotropic-invariant algebras: $\{0\}$, $C_0(H^n)$ and $A(H^n)$, $B(H^n)$, $\overline{A}(H^n)$, $\overline{B}(H^n)$. Only four algebras on the Heisenberg group are invariant under two-sided translations: $\{0\}$, $B(H^n)$, $\overline{B}(H^n)$, $C_0(H^n)$.

Going over to the bounded representation of the Siegel domain as the unit ball $B^n \subset \mathbb{C}^n$, we obtain the description of the closed subalgebras of $C(S)$ on the unit sphere in $\mathbb{C}^n$ that are invariant under analytic automorphisms of the ball that leave a given point of the boundary fixed. In particular, we get a complete list of the closed Möbius-invariant subalgebras of $C(S)$. I established this result previously by a different method ([2], see [9] for details). Here we present a stronger classification—that of the closed subspaces of $C(S)$ invariant under analytic automorphisms with a given fixed point on the boundary, on the assumption that these subspaces are invariant under certain differential operators on the sphere, namely—the infinitesimal biholomorphic translations of the fixed point. This result contains the Nagel-Rudin description of the closed Möbius-invariant subspaces of $C(S)$ [84] as a special case.

As an application to complex analysis in the ball, we present a characterization of boundary values of holomorphic functions in the ball in terms of analytic continuation into sections of the ball by complex lines tangent to orbits of the Heisenberg group.

2.1. Invariant subspaces of $C_0(H^n)$.

2.1.1. Let π denote the regular unitary representation of H^n in the space $L^2(H^n)$ (with respect to Haar measure $dxd\overline{\zeta}d\zeta$) by left translation operators:

$$(\pi_{(x',\zeta')}f)(x,\zeta) = f(x - x' + 2\operatorname{Im}\langle\zeta,\zeta'\rangle, \zeta - \zeta'). \tag{1}$$

A space (resp. algebra) Y of functions on H^n is called a π-space (resp. π-algebra), if $\pi_{(x,\zeta)}f \in Y$ for every $f \in Y$ and every $(x,\zeta) \in H^n$.

LEMMA. *Let Y be a closed π-subspace of $C_0(H^n)$ and Z a π-invariant subset of $Y \cap L^1(H^n)$ which is C-dense in Y. Set $E = \mathrm{cl}_{L^2}(Z)$. Then*

$$Y = \mathrm{cl}_{C_0}[E \cap C_0(H^n)].$$

This follows easily from the fact that every finite Borel measure on H^n that annihilates Y may be approximated in the weak topology (using convolutions with continuous functions) by absolutely continuous measures (with bounded Radon-Nikodym derivatives) that also annihilate Y.

2.1.2. Let F^λ be the Fourier transform operator with respect to the variable x:

$$\left(F^\lambda\right)(\zeta) = f^\lambda(\zeta) = \int_{\mathbb{R}} e^{-i\lambda x} f(x, \zeta)\, dx.$$

The translation operators $\pi_{(x', \zeta')}$ generate the following operators $\pi^\lambda_{(x', \zeta')}$ in the space of functions f^λ:

$$\begin{aligned}\left(\pi^\lambda_{(x', \zeta')} f^\lambda\right)(\zeta) &= \left[\pi_{(x', \zeta')} f\right]^\lambda(\zeta) \\ &= \exp\left[-i\lambda x' + 2i\lambda \operatorname{Im}\langle \zeta, \zeta'\rangle\right] \cdot f^\lambda(\zeta - \zeta'). \qquad (1)\end{aligned}$$

These operators define unitary representation, called π^λ, in $L^2(\mathbb{C}^{n-1})$ (see [21], [22]).

Lemma. *Let E_1, E_2 be two closed π-invariant subspaces of $L^2(H^n)$. Let Z_i be π-invariant subset of $E_i \cap L^1(H^n)$ which is L^2-dense in E_i for $i = 1, 2$. Define $E_i^\lambda = \mathrm{cl}_{L^2}[F^\lambda(Z_i) \cap L^2(\mathbb{C}^{n-1})]$, $\lambda \in \mathbb{R}$. If $E_1^\lambda \subset E_2^\lambda$ for almost all λ, then $E_1 \subset E_2$.*

Proof. It is sufficient to check that if $h \in L^2(H^n)$, $h \perp E_2$, then $h \perp E_1$. Let $f \in Z_2$, $x' \in \mathbb{R}$. By the Parseval equality,

$$\int_{\mathbb{R}\times\mathbb{C}^{n-1}} e^{-i\lambda x'} f^\lambda(\zeta)\overline{h^\lambda(\zeta)}\, d\lambda\, d\overline{\zeta}\, d\zeta = \int_{H^n} (\pi_{(x', 0)} f) \cdot \overline{h}\, dx\, d\overline{\zeta}\, d\zeta.$$

The integral on the right vanishes, because $\pi_{(x', 0)} f \in E_2$ and $h \perp E_2$. Since x' is arbitrary, this implies

$$\int_{\mathbb{C}^{n-1}} f^\lambda(\zeta)\overline{h^\lambda(\zeta)}\, d\overline{\zeta}\, d\zeta = 0 \qquad (2)$$

for almost all λ.

Since f^λ is an arbitrary element of $F^\lambda(Z_2) \cap L^2(\mathbb{C}^{n-1})$, formula (2) is also true for any element f^λ of the closure of E_2^λ. By assumption, it remains true for $g \in Z_1$. Taking f equal to g in (2) and integrating the resulting equality with respect to λ, we obtain $h \perp g$. Thus $h \perp E_1$, and this proves the lemma.

2.1.3. Later we shall need a description of the algebras $A(H^n)$, $B(H^n)$ in terms of the Fourier transform. These algebras contain an approximate unit

$$\gamma_m(x, \zeta) = \frac{(-1)^n m^{4n}}{(x + i|\zeta|^2 + m^2 i)^{2n}}. \qquad (1)$$

Since $\gamma_m \in L^1(H^n)$, each of these algebras contains a dense subset of integrable functions.

The next lemma follows from the Paley-Wiener theorem.

LEMMA. *Let B^λ be the L^2-closure of $B(H^n) \cap L^1(H^n)$. Then $B^\lambda = \{0\}$ if $\lambda < 0$, and $B^\lambda = L^2(\mathbb{C}^{n-1})$ if $\lambda > 0$.*

REMARK. The operator F^λ maps $L^2(H^n)$ into $L^2(\mathbb{C}^{n-1})$ not for all values of λ but only for almost all λ. Therefore B^λ is also defined for almost all λ. To avoid complicating the exposition, we will ignore this point here and in similar constructions below.

2.1.4. Let $\mathscr{E}^\lambda$, $\lambda > 0$, denote the Bergmann space, i.e., the set of all functions $f \in L^2(\mathbb{C}^{n-1})$ such that $e^{\lambda|\zeta|^2} f(\zeta)$ is holomorphic in $\mathbb{C}^{n-1}$. The next lemma follows from the Paley-Wiener theorem for Siegel domains [37].

LEMMA. *Let A^λ be the L^2-closure of $F^\lambda[A(H^n) \cap L^1(H^n)]$. Then $A^\lambda = 0$ if $\lambda < 0$, and $A^\lambda = \mathscr{E}^\lambda$ if $\lambda > 0$.*

2.1.5. For the reader's convenience, we combine the preceding lemmas in one proposition.

PROPOSITION. *Let Y be a closed π-subspace of $C_0(H^n)$ and Z a π-invariant subset of $Y \cap L^1(H^n)$ which is C-dense in Y. Denote*

$$E^\lambda = \mathrm{cl}_{L^2}[F^\lambda(Z) \cap L^2(\mathbb{C}^{n-1})].$$

Then:

(1) *if $E^\lambda = \{0\}$ for a.e. $\lambda < 0$ and $E^\lambda = L^2(\mathbb{C}^{n-1})$ for a.e $\lambda > 0$, then $Y = B(H^n)$;*

(2) *if $E^\lambda = \{0\}$ for a.e $\lambda < 0$ and $E^\lambda = \mathscr{E}^\lambda$ for a.e. $\lambda > 0$, then $Y = A(H^n)$;*

(3) *if $E^\lambda = L^2(\mathbb{C}^{n-1})$ for a.e. λ, then $Y = C_0(H^n)$.*

2.2. π^λ-Subspaces of $L^2(\mathbb{C}^{n-1})$.

2.2.1. Fix some $\lambda > 0$. Consider the unitary representation π^λ of H^n in the space $L^2(\mathbb{C}^{n-1})$ given by formula (1), §2.1.2. Define a λ-convolution by

$$f *^\lambda g = \int_{\mathbb{C}^{n-1}} (\pi^\lambda_{(0,\zeta)} f) \cdot g \, d\bar{\zeta} \, d\zeta. \tag{1}$$

The representation π^λ can be factored into irreducible representations as follows (see [85]).

Consider the functions

$$w_{\mu,\nu,\lambda}(\zeta) = c_{\mu,\nu,\lambda} e^{-\lambda|\zeta|^2} D^\mu_{\bar{u}} \left\{ e^{-2\lambda\langle\zeta,u\rangle} \left[\sqrt{2\lambda}(\bar{\zeta} + \bar{u})\right]^\nu \right\}\Big|_{u=0}, \tag{2}$$

where $\mu, \nu \in \mathbb{Z}_+^{n-1}$ are multi-indices. The functions $w_{\mu,\nu,\lambda}$ belong to $C_0(\mathbb{C}^{n-1}) \cap L^1(\mathbb{C}^{n-1}) \subset L^2(\mathbb{C}^{n-1})$ and constitute an orthonormal basis in $L^2(\mathbb{C}^{n-1})$, provided the normalizing factors $c_{\mu,\nu,\lambda}$ are suitably chosen.

These functions are related to the generalized Laguerre polynomials in the following way:

$$w_{\mu,\nu,\lambda} = c_{\mu,\nu,\lambda} \prod_{k=1}^{n-1} w_{\mu_k,\nu_k,\lambda}(\zeta_k),$$

where

$$w_{\mu_k,\nu_k,\lambda}(\zeta_k) = \begin{cases} e^{-\lambda|\zeta_k|^2} \zeta_k^{\mu_k-\nu_k} L_{\nu_k}^{(\mu_k-\nu_k)}\left(2\lambda|\zeta_k|^2\right), & \mu_k \geq \nu_k, \\ e^{-\lambda|\zeta_k|^2} \overline{\zeta}_k^{\nu_k-\mu_k} L_{\mu_k}^{(\nu_k-\mu_k)}\left(2\lambda|\zeta_k|^2\right), & \mu_k < \nu_k. \end{cases} \tag{3}$$

The explicit formula is

$$w_{\mu,\nu,\lambda} = c_{\mu,\nu,\lambda} e^{-\lambda|\zeta|^2} \times \sum_{0\leq k_j \leq \min(\mu_j,\nu_j)} \frac{(-1)^k \mu!\nu!}{k!(\mu-k)!(\nu-k)!} \left(\sqrt{2\lambda}\overline{\zeta}\right)^{\nu-k} \left(\sqrt{2\lambda}\zeta\right)^{\mu-k}. \tag{4}$$

The following formula relates the basis $w_{\mu,\nu,\lambda}$ to the representation π^λ:

$$w_{\mu,\nu,\lambda} \overset{\lambda}{*} w_{\mu',\nu',\lambda} = \delta_{\nu,\mu'} \cdot w_{\mu,\nu',\lambda}. \tag{5}$$

Note also the following relation, which follows from (2):

$$w_{\mu,\nu,\lambda} = c_{\mu,\nu,\lambda} D_{\overline{u}}^{\mu} \left[e^{-\lambda|u|^2} \pi^\lambda_{(0,-u)} w_{0,\nu,\lambda} \right] \Big|_{u=0}. \tag{6}$$

REMARK. The positive factors $c_{\mu,\nu,\lambda}$ in the above formulas may be different, but this is not essential for what follows.

Let W_ν^λ denote the closed linear span in $L^2(\mathbb{C}^{n-1})$ of the set of functions

$$\pi^\lambda_{(0,u)} w_{0,\nu,\lambda}(\zeta) = c_{0,\nu,\lambda} \exp\left[-\lambda|u|^2 - \lambda|\zeta|^2 + 2\lambda\langle\zeta, u\rangle\right] \cdot \left[\sqrt{2\lambda}(\zeta - u)\right]^\nu,$$

where u is an arbitrary element of $\mathbb{C}^{n-1}$.

LEMMA. *The space W_ν^λ is a direct sum of one-dimensional spaces*:

$$W_\nu^\lambda = \bigoplus_{\mu\in\mathbb{Z}_+^{n-1}} \mathbb{C} \cdot w_{\mu,\nu,\lambda}.$$

PROOF. It follows from the equality $w_{\mu,\nu,\lambda} = w_{\mu,0,\lambda} \overset{\lambda}{*} w_{0,\nu,\lambda}$ that $w_{\mu,\nu,\lambda} \in W_\nu^\lambda$ for any $\mu \in \mathbb{Z}_+^{n-1}$, and hence

$$\bigoplus_{\mu\in\mathbb{Z}_+^{n-1}} \mathbb{C} \cdot w_{\mu,\nu,\lambda} \subset W_\nu^\lambda.$$

The reverse inclusion follows from the Taylor expansion

$$\pi^\lambda_{(0,u)} w_{0,\nu,\lambda}(\zeta) = c_{0,\nu,\lambda} e^{-\lambda|u|^2} \sum_{\mu\in\mathbb{Z}_+^{n-1}} c^{-1}_{\mu,\nu,\lambda} \frac{\overline{u}^\mu}{\mu!} w_{\mu,\nu,\lambda}(\zeta).$$

2.2.2. We call a space E of functions in $\mathbb{C}^{n-1}$ *a polydisk space if the function* $\zeta \to f(\lambda_1\zeta_1, \dots, \lambda_{n-1}\zeta_{n-1})$ *belongs to* E *for all* $f \in E$ *and all complex numbers* λ_k *such that* $|\lambda_k| = 1$.

LEMMA. *Let* E *be a closed polydisk subspace of* $L^2(\mathbb{C}^{n-1})$ *which is invariant with respect to the representation* π^λ, $\lambda > 0$. *Then* $W_\nu^\lambda \subset E$ *if and only if*

$$\int_{\mathbb{C}^{n-1}} f(\zeta)\overline{w_{\mu,\nu,\lambda}(\zeta)}\, d\overline{\zeta}\, d\zeta \tag{1}$$

for some $f \in E$ *and* $\mu \in \mathbb{Z}_+^{n-1}$.

Henceforth, spaces invariant with respect to π^λ will be called π^λ-spaces.

PROOF. Let $f \in E$ and $\mu \in \mathbb{Z}_+^{n-1}$ satisfy condition (1), and let

$$f = \sum_{\alpha,\beta\in\mathbb{Z}_+^{n-1}} a_{\alpha,\beta} w_{\alpha,\beta,\lambda}$$

be the orthogonal expansion of f with respect to the λ-basis. Define g by

$$g = w_{\nu,\mu,\lambda} \overset{\lambda}{*} f = \sum_{\beta\in\mathbb{Z}_+^{n-1}} a_{\mu,\beta} w_{\nu,\beta,\lambda}.$$

The second formula for g follows from formula (5) in §2.2.1. Clearly, $g \in E$.

Consider the radialization operator

$$(Rg)(\zeta) = \frac{1}{(2\pi)^{n-1}} \int_{(0,2\pi)^{n-1}} g\left(e^{i\varphi_1}\zeta_1, \dots, e^{i\varphi_{n-1}}\zeta_{n-1}\right) d\varphi_1 \cdots d\varphi_{n-1},$$

which maps E into E. Since the functions $w_{\nu,\beta,\lambda}$ are homogeneous (see formula (3) in §2.2.4), we have $Rw_{\nu,\beta,\lambda} = 0$ for $\nu \neq \beta$. Therefore $Rg = a_{\mu,\nu} w_{\nu,\nu,\lambda}$. By assumption $a_{\mu,\nu} \neq 0$; hence,

$$w_{0,\nu,\lambda} = a_{\mu,\nu}^{-1}\left(w_{0,\nu,\lambda} \overset{\lambda}{*} Rg\right) \in E.$$

But then $W_\nu^\lambda \subset E$.

2.2.3. For negative λ define the space W_ν^λ as the complex conjugate of $W_\nu^{-\lambda}$. Since $\overline{w_{\mu,\nu,|\lambda|}} = w_{\mu,\nu,|\lambda|}$, Lemma 2.2.1 implies that

$$W_\nu^\lambda = \bigoplus_{\mu\in\mathbb{Z}_+^{n-1}} \mathbb{C}\cdot w_{\nu,\mu,|\lambda|}$$

for λ negative.

LEMMA. *Let $\lambda \neq 0$. Every closed polydisk π^λ-subspace $E \subset L^2(\mathbb{C}^{n-1})$, $E \neq \{0\}$, can be represented as a direct sum*

$$E = \bigoplus_{\nu \in \Delta} W_\nu^\lambda,$$

where Δ is a subset of $\mathbb{Z}_+^{n-1}$.

PROOF. First let $\lambda > 0$. Define Δ as the set of all $\nu \in \mathbb{Z}_+^{n-1}$ such that $W_\nu^\lambda \subset E$. Let $f \in E$, $f = \sum_{\mu,\nu \in \mathbb{Z}_+^{n-1}} a_{\mu,\nu} w_{\mu,\nu,\lambda}$. By Lemma 2.2.2, $a_{\mu,\nu} \neq 0$ only if $\nu \in \Delta$. Thus $E \subset \bigoplus_{\nu\in\Delta} W_\nu^\lambda$. The reverse inclusion is also true. The case $\lambda < 0$ is reduced to the previous one by complex conjugation.

2.2.4. A function space Y on the group H^n is *isotropic* if $f \circ \delta_t \in Y$ for any $f \in Y$ and any anisotropic dilatation $\delta_t(x, \zeta) = (t^2 x, t\zeta)$, $t > 0$. Y is called *a polydisk space* if the function $(x, \zeta) \to f(x, \lambda_1\zeta_1, \dots, \lambda_{n-1}\zeta_{n-1})$ belongs to Y for any $f \in Y$ and any $\lambda_k \in \mathbb{C}$, $|\lambda_k| = 1$.

LEMMA. *Let Y be a closed polydisk π-subspace of $C_0(\mathbb{C}^{n-1})$ and Z a π-invariant subset of $Y \cap L^1(H^n)$ which is C-dense in Y. Define*

$$E^\lambda = \mathrm{cl}_{L^2}\left[F^\lambda(Z) \cap L^2\left(\mathbb{C}^{n-1}\right)\right], \qquad \lambda \neq 0.$$

The space E^λ satisfies the assumptions of Lemma 2.2.3; *therefore,*

$$E^\lambda = \bigoplus_{\nu \in \Delta^\lambda} W_\nu^\lambda$$

for some $\Delta^\lambda \subset \mathbb{Z}_+^{n-1}$. Suppose that Y is anisotropic. Then $\Delta^\lambda = \Delta^{\lambda'}$ for any λ, λ' such that $\lambda\lambda' > 0$.

PROOF. We have the following relation between Fourier transforms and the dilatations δ_t:

$$F^\lambda(f \circ \delta_t)(\zeta) = t^{-2}\left(F^{t^{-2}\lambda} f\right)(t\zeta). \tag{1}$$

It is enough to verify the claim for $\lambda, \lambda' > 0$. Let $\nu \in \Delta^\lambda$. Then $w_{0,\nu,\lambda} \in E^\lambda$, and by the definition of E^λ there exists a sequence $f_m \in Z$ converging to $w_{0,\nu,\lambda}$ in the norm of $L^2(\mathbb{C}^{n-1})$. Set $g_m = t^2 f_m \circ \delta_t$, where $t = (\lambda'/\lambda)^{1/2}$. By (1),

$$g_m^{\lambda'}(\zeta) = f_m^\lambda(t\zeta),$$

and therefore the sequence g_m, which is of course contained in $F^{\lambda'}(Z)$, converges to $w_{0,\nu,\lambda}(t\zeta)$ in $L^2(\mathbb{C}^{n-1})$. But

$$w_{0,\nu,\lambda}(t\zeta) = c_{0,\nu,\lambda}\left(\sqrt{2\lambda}t = c_{0,\nu,\lambda}\left(\sqrt{2\lambda'\zeta}\right)^\nu \bar\zeta\right)^\nu,$$

and so $w_{0,\nu,\lambda'} \in E^{\lambda'}$, which means that $\nu \in \Delta^{\lambda'}$. Thus $\Delta^{\lambda} \subset \Delta^{\lambda'}$. Interchanging λ and λ', we obtain the reverse inclusion.

2.2.5.* An example of the application of Fourier analysis on the Heisenberg group to the complex analysis is the characterization of the boundary values of holomorphic functions in the unit ball of $\mathbb{C}^n$.

Let $A(S)$ denote the algebra of boundary values on the unit sphere $S \subset \mathbb{C}^n$ of functions holomorphic in the open unit ball B^n and continuous in $B^n \cup S$. Let Q_t, $t > 0$, denote the ellipsoid

$$Q_t = \left\{ w \in \mathbb{C}^n : 1 - |w|^2 = t^2 |1 - w_1|^2 \right\}$$

which is contained in the unit ball B^n and is tangent to the unit sphere at the point $e = (1, 0, \dots, 0) \in S$. Let Γ denote the family of complex straight lines tangent to Q_t.

THEOREM. *Let $f \in C(S)$ be such that, for any complex straight line $\Lambda \in \Gamma$, the restriction $f|_{\Lambda \cap S}$ to the circle $\Lambda \cap S$ may be holomorphically extended into the disk $\Lambda \cap B^n$. Then $f \in A(S)$.*

PROOF. Replacing f by $f - f(e)$, we may assume $f(e) = 0$. Let $Y(S)$ denote the space of all $f \in C(S)$, $f(e) = 0$, that satisfy the assumptions of the theorem. The Cayley transformation $\Phi : B^n \to \Omega^n$ maps Q_t onto the set

$$\Phi(Q_t) = \left\{ (z, \zeta) \in \mathbb{C} \times \mathbb{C}^{n-1} : \operatorname{Im} z = |\zeta|^2 + t^2 \right\},$$

which is the orbit of the point $(it, 0) \in \Omega^n$ under the action of the group H^n, considered as a subgroup of $\operatorname{Aut}(\Omega^n)$. The Cayley transformation maps Γ onto the family $\Phi(\Gamma)$ of complex straight lines tangent to $\Phi(Q_t)$.

Define

$$Y(\partial\Omega^n) = \left\{ f \circ \Phi^{-1} : f \in Y(S) \right\},$$

and let Y be the space of functions defined on H^n that can be represented as

$$f(x, \zeta) = g\left(x + i|\zeta|^2, \zeta\right), \qquad g \in Y(\partial\Omega^n).$$

Then Y is a closed π-subspace of $C_0(H^n)$ and $Y \cap L^1(H^n)$ is C-dense in Y, because Y is an algebra and contains an approximate unit γ_m (cf. formula (1), §2.1.3). It is also clear that Y is invariant under unitary coordinate transformations; in particular, Y is a polydisk space.

Define

$$E^{\lambda} = \operatorname{cl}_{L^2} \left\{ F^{\lambda} \left[Y \cap L^1(H^n) \right] \cap L^2\left(\mathbb{C}^{n-1}\right) \right\}, \qquad \lambda \neq 0.$$

**Remark for the English edition*: For further development of the results of this section see [112]–[114].

The result of §2.2.3 implies that $E^\lambda = \bigoplus_{\nu\in\Delta^\lambda} W_\nu^\lambda$. We claim that $\Delta^\lambda = \{0\}$ for $\lambda > 0$, i.e., $E^\lambda = W_0^\lambda = \mathscr{E}^\lambda$. Indeed, consider the complex straight line $\Lambda \in \Gamma$ defined by

$$w_1 = (t^2-1)/(t^2+1), \qquad w_2 = \cdots = w_n.$$

The image $\Phi(\Lambda)$ of this line under the Cayley transformation is the complex straight line $z = it^2$, $\zeta_1 = \cdots = \zeta_n$. By the definition of Y, the restriction of any function $f \in Y \cap L^1(H^n)$ to the circle

$$x = 0, \quad \zeta_1 = \cdots = \zeta_{n-1} = \rho e^{i\theta}, \quad \rho = \frac{t}{\sqrt{n-1}}, \qquad 0 < \theta < 2\pi,$$

may be continued analytically into the disk

$$x = 0, \quad \zeta_1 = \cdots = \zeta_{n-1} = \rho\eta, \quad |\eta| < 1.$$

The functions $\pi_{(-x',0)}f$, $x' \in \mathbb{R}$, clearly have the same property, and therefore

$$\int_0^{2\pi} f\left(x', \rho e^{i\theta}, \ldots, \rho e^{i\theta}\right) e^{ik\theta}\, d\theta = 0, \qquad k = 1, 2, \ldots.$$

Multiplying both sides of this equality by $e^{-i\lambda x'}$ and integrating with respect to x', we get

$$\int_0^{2\pi} f^\lambda\left(\rho e^{i\theta}, \ldots, \rho e^{i\theta}\right) e^{ik\theta}\, d\theta = 0. \tag{1}$$

Since we can take $\pi_{(0,\zeta')}f$ instead of f, it follows that (1) is also true for functions $\pi_{(0,\zeta')}f$, $\zeta' \in \mathbb{C}^{n-1}$, hence also for λ-convolutions $w_{0,\nu,\lambda} *^\lambda f^\lambda$. Therefore, we can replace f^λ in (1) by $w_{0,\nu,\lambda}$, where $\nu \in \Delta^\lambda$. Indeed, $w_{0,\nu,\lambda}$ is the L^2-limit of a sequence f_m^λ, $f_m \in Y \cap L^1(H^n)$. By the estimate for the uniform norm

$$\|w_{0,0,\lambda} *^\lambda w_{0,\nu,\lambda} - w_{0,0,\lambda} *^\lambda f_m^\lambda\| \le \|w_{0,0,\lambda}\|_2 \cdot \|w_{0,\nu,\lambda}\|_2$$

the sequence $w_{0,0,\lambda} *^\lambda f_m^\lambda$ converges uniformly to $w_{0,\nu,\lambda} = w_{0,0,\lambda} *^\lambda w_{0,\nu,\lambda}$ as $m \to \infty$, hence, (1) also holds for the limit function $w_{0,\nu,\lambda}$. But this is possible only if $\nu = 0$, because

$$w_{0,\nu,\lambda}\left(\rho e^{i\theta}, \ldots, \rho e^{i\theta}\right) = c_{0,\nu,\lambda}(2\lambda)^{|\nu|/2} e^{-i|\nu|\theta}.$$

Thus we have proved that $\Delta^\lambda = \{0\}$ and $E^\lambda = \mathscr{E}^\lambda$ if $\lambda > 0$.

Now suppose $\lambda < 0$. If E^λ is not trivial, it must contain $w_{\nu,\mu,|\lambda|}$ for some $\nu \in \mathbb{Z}_+^{n-1}$ and every $\mu \in \mathbb{Z}_+^{n-1}$. As in the case $\lambda > 0$, this implies that $w_{\nu,\mu,|\lambda|}$ must satisfy condition (1). Take $\mu \in \mathbb{Z}_+^{n-1}$ with $\mu_j > \nu_j$, $j = 1, \ldots, n-1$; put $k = (\mu_1 - \nu_1) + \cdots + \left(\mu_{n-1} - \nu_{n-1}\right)$ in (1); and replace

f^λ by $w_{\nu,\mu,|\lambda|}$. Using formula (3) of §2.2.1 for the function $w_{\nu,\mu,,|\lambda|}$, we conclude from (1) that

$$\prod_{j=1}^{n-1} L_{\nu_j}^{(\mu_j-\nu_j)}\left(2\lambda\rho^2\right) = 0. \tag{2}$$

Note that the only condition imposed on the indices is $\mu_j > \nu_j$. But by Markov's theorem [27] the roots of the Laguerre polynomials $L_m^{(\alpha)}$ are strictly increasing functions of α, so (2) cannot hold for any μ_j greater than ν_j. Therefore, $E^\lambda = \{0\}$ for $\lambda < 0$. It follows from Proposition 2.1.5 that $Y = A(H^n)$ and accordingly $Y(S) = A(S)$.

REMARK. The ellipsoid Q_t is an orbit of the Heisenberg group, considered as a subgroup of $\mathrm{Aut}(B^n)$. Nagel and Rudin [84] gave an analogous characterization of the functions in $A(S)$, with Q_t an orbit of the unitary subgroup $U(n) \subset \mathrm{Aut}(B^n)$, i.e., a sphere tS, $0 < t < 1$.

2.3. Invariant subalgebras of $C_0(H^n)$.

2.3.1. Let Y be a closed π-subspace of $C_0(H^n)$. Define the spectrum $\sigma(Y)$ of Y as the set of common zeros of the Fourier transforms of functions $h \in L^1(\mathbb{R})$ such that

$$\int_{\mathbb{R}} f(x-t, \zeta)h(t)\,dt = 0$$

for any $f \in Y$ and all $x \in \mathbb{R}$, $\zeta \in \mathbb{C}$.

One can associate with Y a translation-invariant space of $C_0(\mathbb{C}^{n-1})$-valued functions $x \to f(x, \cdot)$ defined on the real axis. Using the methods developed in [81] to study translation-invariant spaces of scalar functions on the real axis without essential changes, we can prove the following

PROPOSITION. (1) *If Y is an algebra, then $\sigma(Y)$ is a semigroup.*

(2) *If Y contains an approximate unit, then $0 \in \sigma(Y)$.*

(3) *Suppose there exists a subset $Z \subset Y \cap L^1(H^n)$ which is C-dense in Y. Then $\sigma(Y)$ is the closure of the open semigroup*

$$\sigma_0(Y) = \left\{\lambda : F^\lambda(Z) \neq \{0\}\right\}.$$

2.3.2. Let us now describe the two-sided invariant algebras on H^n.

THEOREM. *Let Y be a closed subalgebra of $C_0(H^n)$, invariant under right and left translations, which contains a dense subset of integrable functions and an approximate unit. Then Y is one of the four algebras in the following diagram (the arrows represent inclusions):*

$$\begin{array}{ccc} \{0\} & \longrightarrow & B(H^n) \\ \downarrow & & \downarrow \\ \overline{B}(H^n) & \longrightarrow & C_0(H^n) \end{array}$$

PROOF. Define

$$E^{\lambda} = \mathrm{cl}_{L^2}\{F^{\lambda}[Y \cap L^1(H^n)] \cap L^2(\mathbb{C}^{n-1})\},$$

where $\lambda \neq 0$. Since Y is two-sided invariant, E^{λ} is invariant under λ-convolutions from the left and from the right by the elements of $L^1(\mathbb{C}^{n-1})$.

Let $\lambda > 0$, and suppose $E^{\lambda} \neq \{0\}$. Consider a nonzero function $f \in E^{\lambda}$, and expand it in Fourier series with respect to the λ-basis:

$$f = \sum_{\alpha, \beta \in \mathbb{Z}_+^{n-1}} a_{\alpha, \beta} w_{\alpha, \beta, \lambda}.$$

Let $a_{\alpha_0, \beta_0} \neq 0$. By formula (5) of §2.2.1,

$$w_{\mu, \nu, \lambda} = \frac{1}{a_{\alpha_0, \beta_0}} \left(w_{\mu, \alpha_0} \overset{\lambda}{*} f \overset{\lambda}{*} w_{\beta_0, \nu, \lambda} \right),$$

and therefore $w_{\mu, \nu, \lambda} \in E^{\lambda}$ for any $\mu, \nu \in \mathbb{Z}_+^{n-1}$. Thus the assumption $E^{\lambda} \neq \{0\}$ implies $E^{\lambda} = L^2(\mathbb{C}^{n-1})$.

Substituting $\overline{Y}$ for Y in this argument, we see that, if $\lambda < 0$, then E^{λ} is either $\{0\}$ or $L^2(\mathbb{C}^{n-1})$.

The set $\sigma(Y)$ defined in Section 4.1 coincides up to a set of measure zero with the set of all λ such that $E^{\lambda} \neq \{0\}$. Since $\sigma(Y)$ is a subsemigroup of $\mathbb{R}$ that contains zero and coincides with the closure of its interior (see Proposition 2.3.1), it follows that $\sigma(Y)$ is one the following three semigroups: $\mathbb{R}_+$, $\mathbb{R}_-$, $\mathbb{R}$. The proof is completed by applying Proposition 2.1.5.

2.3.3. We will now describe the left invariant subalgebras of $C_0(H^n)$. We will see that this class is far larger than that of two-sided invariant algebras.

In this and following sections Y will be a closed π-subalgebra of $C_0(H^n)$ such that $Y \cap L^1(H^n)$ is dense in the uniform norm in Y.

In order to obtain a legitimate definition of multiplication for Fourier transforms, we single out a special dense subset of Y.

LEMMA. *Let Y^0 be the set of all $f \in Y \cap L^1(H^{n-1})$ such that $f^{\lambda} \in C_0(H^n)$ for almost all λ and the uniform norms of f^{λ} are bounded:* $\operatorname{ess\,sup}_{\lambda} \|f^{\lambda}\|_{\infty} < \infty$. *Then Y^0 is uniformly dense in Y.*

PROOF. Let $f \in Y \cap L^1(H^n)$ and φ be a continuous function with compact support defined in $\mathbb{C}^{n-1}$. Define

$$\pi_{\varphi} f = \int_{\mathbb{C}^{n-1}} \varphi(u) \pi_{(0, u)} f \, d\bar{u} \, du.$$

The function $\pi_{\varphi} f$ belongs to $Y \cap L^1(H^n)$, and

$$\left(\pi_{\varphi} f\right)^{\lambda}(\zeta) = \int_{\mathbb{C}^{n-1}} \varphi(\zeta - \eta) \exp[2i\lambda \operatorname{Im}\langle \zeta, \zeta - \eta \rangle] f^{\lambda}(\eta) \, d\bar{\eta} \, d\eta.$$

Since $f^\lambda \in L^1(\mathbb{C}^{n-1})$ for almost all λ, it follows from the Lebesgue dominated convergence theorem that $(\pi_\varphi f)^\lambda \in C_0(\mathbb{C}^{n-1})$. In addition,

$$\|(\pi_\varphi f)^\lambda\|_\infty \leq \|\varphi\|_\infty \cdot \|f\|_{L^1}.$$

Therefore, $\pi_\varphi f \in Y^0$. It remains to observe that the functions $\pi_\varphi f$ constitute a C-dense subset of $Y \cap L^1(H^n)$, which is in turn dense in Y.

2.3.4. LEMMA. *Define*

$$E^\lambda = \mathrm{cl}_{L^2}[F^\lambda(Y^0) \cap L^2(\mathbb{C}^{n-1})].$$

Let $\alpha \in E^{\lambda_1} \cap C_0(\mathbb{C}^{n-1})$, $\beta \in E^{\lambda_2} \cap C_0(\mathbb{C}^{n-1})$. *Then* $\alpha\beta \in E^{\lambda_1+\lambda_2}$.

PROOF. Suppose first that $\alpha \in F^{\lambda_1}(Y^0)$, $\beta \in F^{\lambda_2}(Y^0)$. This means that $\alpha = f^{\lambda_1}$, $\beta = g^{\lambda_2}$, where $f, g \in Y^0$. Take a sequence $\delta_m \in L^1(\mathbb{R}^n)$ such that the Fourier transforms $\hat{\delta}_m$ form a δ-sequence of positive functions with compact support. For example, set

$$\delta_m(x) = \frac{1}{2\pi}\frac{m^2}{x^2}\sin^2\frac{x}{m}.$$

Now define

$$g_m(x, \zeta) = \int_{\mathbb{R}} e^{i\lambda_2 x'}\delta_m(x')g(x - x', \zeta)\,dx',$$

and let $h_m = fg_m$. Then $h_m \in Y \cap L^1(H^n)$ and

$$h_m^{\lambda_1+\lambda_2}(\zeta) = \int_{\mathbb{R}} f^{\lambda_1+\lambda_2-\tau}(\zeta)\hat{\delta}_m(\tau - \lambda_2)g^\tau(\zeta)\,d\tau = \int_{\mathbb{R}} f^{\lambda_1-\tau}(\zeta)\hat{\delta}_m(\tau)g^{\lambda_2+\tau}\,d\tau.$$

Taking into account that $\int_{\mathbb{R}} \hat{\delta}_m(\tau)\,d\tau = 1$, we get

$$\alpha(\zeta)\beta(\zeta) - h_m^{\lambda_1+\lambda_2}(\zeta) = \int_{\mathbb{R}} \hat{\delta}_m(\tau)\left[\alpha(\zeta)\beta(\zeta) - f^{\lambda_1-\tau}(\zeta)g^{\lambda_2+\tau}(\zeta)\right]d\tau,$$

and hence

$$\|\alpha\beta - h_m^{\lambda_1+\lambda_2}\|_2 \leq \int_{\mathbb{R}} \hat{\delta}_m(\tau)s(\tau)\,d\tau, \tag{1}$$

where $s(\tau) = \|\alpha\beta - f^{\lambda_1-\tau}g^{\lambda_2+\tau}\|_2$. Let us estimate $s(\tau)$:

$$0 \leq s(\tau) \leq \|\alpha\|_\infty\|\beta - g^{\lambda_2+\tau}\|_2 + \|g^{\lambda_2+\tau}\|_\infty \cdot \|\alpha - f^{\lambda_1-\tau}\|_2. \tag{2}$$

We have

$$\|\beta - g^{\lambda_2+\tau}\|_2 = \int_{\mathbb{C}^{n-1}} \left|\int_{\mathbb{R}} (1 - e^{i\tau x})e^{i\lambda_2 x}g(x, \zeta)\,dx\right|^2 d\bar{\zeta}\,d\zeta. \tag{3}$$

Since g is absolutely integrable, the inner integral in (3) tends to zero as $\tau \to 0$ for almost all ζ. In addition,

$$\left| \int_{\mathbb{R}} \left(1 - e^{i\tau x}\right) e^{i\lambda_2 x} g(x, \zeta)|\, dx \right|^2 \leq 4 \sup_{\lambda} \|g^{\lambda}\|_{\infty} \cdot \int_{\mathbb{R}} |g(x, \zeta)|\, dx,$$

where the function of ξ on the right belongs to $L^1(\mathbb{C}^{n-1})$. The dominated convergence theorem now implies that $\lim_{\tau\to 0} \|\beta - f^{\lambda_2+\tau}\|_2 = 0$. An analogous argument proves that $\lim_{\tau\to 0} \|\alpha - f^{\lambda_1-\tau}\|_2 = 0$. By the definition of Y^0,

$$\|\alpha\|_{\infty} = \|f^{\lambda_1}\|_{\infty} < \infty, \qquad \operatorname*{ess\,sup}_{\tau} \|g^{\lambda_2+\tau}\|_{\infty} < \infty,$$

and therefore it follows from (2) that $\lim_{\varepsilon\to 0} \operatorname{ess\,sup}_{|\tau|<\varepsilon} s(\tau) = 0$. Since $\hat{\delta}_m$ is a δ-sequence, this enables us to infer from (1) that

$$\lim_{m\to\infty} \left\| \alpha\beta - h_m^{\lambda_1+\lambda_2} \right\|_2 = 0.$$

But $h_m^{\lambda_1+\lambda_2} \in E^{\lambda_1+\lambda_2}$, and therefore $\alpha\beta \in E^{\lambda_1+\lambda_2}$.

In the general case, there exists a sequence $f_m \in Y^0$ such that

$$\lim_{m\to\infty} \left\| \alpha - f_m^{\lambda_1} \right\|_2 = 0.$$

Let us approximate the function β by a sequence $g_m^{\lambda_2} \in F^{\lambda_2}(Y^0)$, but this time with respect to the uniform norm. This is possible, because β belongs to the space $E^{\lambda_2} \cap C_0(\mathbb{C}^{n-1})$, which is contained in the uniform closure of the set $F^{\lambda_2}(Y^0) \subset C_0(\mathbb{C}^{n-1})$. (That $E^{\lambda_2} \cap C_0(\mathbb{C}^{n-1}) \subset \operatorname{cl}_{C_0(\mathbb{C}^{n-1})} F^{\lambda_2}(Y^0)$ can be proved by an argument similar to that of the proof of Lemma 2.1.1, using λ-convolutions.)

Choosing sequences f_m, $g_m \in Y^0$ as stated, we conclude from the inequality

$$\left\| \alpha\beta - f_m^{\lambda_1} g_m^{\lambda_2} \right\|_2 \leq \|\alpha\|_2 \left\| \beta - g_m^{\lambda_2} \right\|_{\infty} + \left\| g_m^{\lambda_2} \right\|_{\infty} \left\| \alpha - f_m^{\lambda_1} \right\|_2$$

that $\alpha\beta \in E^{\lambda_1+\lambda_2}$, as we have already proved that $f_m^{\lambda_1} g_m^{\lambda_2} \in E^{\lambda_1+\lambda_2}$.

2.3.5. We must now learn to multiply the irreducible subspaces W_{ν}^{λ} introduced in §2.2.1.

Let $W_{\nu}^{\lambda} \odot W_{\nu'}^{\lambda'}$ denote the subspace of $L^2(\mathbb{C}^{n-1})$ defined as the L^2-closure of the set of all finite sums

$$\sum \alpha_i \varphi_i(\zeta)\psi_i(\zeta), \qquad \alpha_i \in \mathbb{C},\ \varphi_i \in W_{\nu}^{\lambda},\ \psi_i \in W_{\nu'}^{\lambda'}.$$

LEMMA. *The space* $W_{\nu}^{\lambda} \odot W_{\nu'}^{\lambda'}$ *may be represented as the direct sum*

$$W_{\nu}^{\lambda} \odot W_{\nu'}^{\lambda'} = \bigoplus_{0\leq k\leq \nu+\nu'} W_k^{\lambda+\lambda'} \tag{1}$$

of the minimal subspaces $W_k^{\lambda+\lambda'}$ *over all* $k \in \mathbb{Z}_+^{n-1}$ *such that* $0 \le k_j \le \nu_j + \nu_j'$, $j = 1, \dots, n-1$.

PROOF. Since $W_\nu^\lambda \odot W_{\nu'}^{\lambda'}$ is a $\pi^{\lambda+\lambda'}$-invariant polydisk space, it follows from Lemma 2.2.3 that it may be represented as the direct sum of the spaces $W_k^{\lambda+\lambda'}$. The inclusion $W_k^{\lambda+\lambda'} \subset W_\nu^\lambda \odot W_{\nu'}^{\lambda'}$ holds, by Lemma 2.2.2, if and only if the integral

$$I_k(u, u', v) = \int_{\mathbb{C}^{n-1}} \pi^\lambda_{(0,u)} w_{0,\nu,\lambda}(\zeta) \cdot \pi^{\lambda'}_{(0,u')} w_{0,\nu',\lambda'}(\zeta) \overline{\pi^{\lambda+\lambda'}_{(0,v)} w_{0,k,\lambda+\lambda'}(\zeta)}\, d\overline{\zeta}\, d\zeta \tag{2}$$

does not vanish for some $u, u', v \in \mathbb{C}^{n-1}$.

Using formula (2), §2.2.1, for the functions $w_{0,\nu\lambda}$, $w_{0,\nu',\lambda'}$, and $w_{0,k,\lambda+\lambda'}$, changing the variable $\zeta \to \zeta + v$, and performing a few elementary calculations, we obtain

$$I_k(u, u', v) = c \int_{\mathbb{C}^{n-1}} F(\zeta) G(\zeta) e^{-2(\lambda+\lambda')|\zeta|^2}\, d\overline{\zeta}\, d\zeta,$$

where

$$F(\zeta) = \zeta^k \exp\left[2\lambda\langle\zeta, u - v\rangle + 2\lambda'\langle\zeta, u' - v\rangle\right],$$

$$G(\zeta) = \left(\overline{\zeta} - \overline{u} + \overline{v}\right)^\nu \left(\overline{\zeta} - \overline{u}' + \overline{v}\right)^{\nu'};$$

the nonzero factor c depends on $\lambda, \lambda', u, u', v$.

Substitute the power series

$$F(\zeta) = \sum_{\alpha \ge k} a_\alpha \zeta^\alpha, \qquad G(\zeta) = \sum_{0 \le \beta \le \nu+\nu'} b_\beta \overline{\zeta}^\beta$$

into the integral. As monomials of different degrees are orthogonal:

$$\int_{\mathbb{C}^{n-1}} \zeta^\alpha \overline{\zeta}^\beta e^{-2(\lambda+\lambda')|\zeta|^2}\, d\overline{\zeta}\, d\zeta = 0, \qquad \alpha \ne \beta,$$

we have $I_k(u, u', v) = 0$, provided there are no monomials of the same degree in the expansions of $F(\zeta)$ and $G(\zeta)$. This will be the case if $k_j > \nu_j + \nu_j'$ for at least one j. This proves the inclusion

$$W_\nu^\lambda \odot W_{\nu'}^{\lambda'} \subset \bigoplus_{0 \le k \le \nu+\nu'} W_k^{\lambda+\lambda'}$$

The proof of the reverse inclusion is more complicated and uses reduction formulas for the Laguerre polynomials. Suppose that for some $k \in \mathbb{Z}_+^{n-1}$, $0 \le k \le \nu + \nu'$, the space $W_k^{\lambda+\lambda'}$ is not contained in $W_\nu^\lambda \odot W_{\nu'}^{\lambda'}$. As already stated, this means that $I_k(u, u', v) = 0$ for all $u, u', v \in \mathbb{C}^{n-1}$. Then all the derivatives $D^{\mu}_{\bar{u}} D^{\mu'}_{\bar{u}'} D^{\mu''}_{v} I_k(u, u', v)$ vanish at the point $u = u' = v = 0$ for all multi-indices μ, μ', μ''.

Differentiating (2) and taking into account formula (6) of §2.2.1, we get

$$\int_{\mathbb{C}^{n-1}} w_{\mu,\nu,\lambda}(\zeta) w_{\mu',\nu',\lambda'}(\zeta) \overline{w_{\mu'',k,\lambda+\lambda'}(\zeta)}\, d\bar{\zeta}\, d\zeta = 0. \tag{3}$$

We now show that (3) does not hold if $\mu = \nu'$, $\mu' = \nu$, $\mu'' = k$.

The integral (3) can be represented as the product of $n-1$ one-dimensional integrals of the same type, since the integrand is the product of factors corresponding to $\zeta_1, \ldots, \zeta_{n-1}$, (see (3), §2.2.1). It will thus suffice to prove the assertion in the case $n = 2$. Without loss of generality one can assume that $\nu' \geq \nu$.

Substitute the representation of the functions $w_{\mu,\nu,\lambda}$ in terms of Laguerre polynomials into (3). Changing the variables, we obtain the equality

$$\int_0^\infty t^{\nu'-\nu} L_\nu^{(\nu'-\nu)}(\gamma t) L_\nu^{(\nu'-\nu)}((1-\gamma)t) L_k^{(0)}(t) e^{-t}\, dt = 0, \tag{4}$$

where $\gamma = \lambda/(\lambda + \lambda')$.

To calculate the integral (4), we use the following formula for products of Laguerre polynomials [17, p. 192]:

$$L_\nu^{(\alpha)}(x) L_\nu^{(\alpha)}(y) = \frac{(\alpha+\nu)!}{\nu!} \sum_{m=0}^{\nu} \frac{1}{m!(\alpha+m)!} (xy)^m L_{\nu-m}^{(\alpha+2m)}(x+y).$$

Putting $x = \gamma t$, $y = (1-\gamma)t$, $\alpha = \nu' - \nu$ and substituting the result into (4), we get

$$\sum_{m=0}^{\nu} \frac{\gamma^m (1-\gamma)^m}{m!(\nu'-\nu+m)!} \int_0^\infty t^{\nu'-\nu+2m} L_{\nu-m}^{(\nu'-\nu+2m)}(t) L_k^{(0)}(t) e^{-t}\, dt = 0. \tag{5}$$

In order to equate the superscripts of the Laguerre polynomials in the integrand, we express $L_k^{(0)}$ in terms of $L_p^{(\nu'-\nu+2m)}$ (see [17, p. 192]):

$$L_k^{(0)}(t) = \sum_{j=0}^{\nu'-\nu+2m} \frac{(-1)^j}{j!} \binom{\nu'-\nu+2m}{j} L_{k-j}^{(\nu'-\nu+2M)}(t). \tag{6}$$

Laguerre polynomials with the same superscripts satisfy the orthogonality conditions

$$\int_0^\infty L_{k-j}^{(\nu'-\nu+2m)}(t) L_{\nu-m}^{(\nu'-\nu+2m)}(t) t^{\nu'-\nu+2m} e^{-t}\, dt = \delta_{k-j,\nu-m},$$

and, therefore, substituting (6) into (5), we obtain

$$\sum_{m=m_0}^{\nu} (-1)^{m-m_0} a_m z^m = 0, \tag{7}$$

where we have used the notation

$$z=\gamma(1-\gamma),\quad m_0=\max(\nu-k,\,k-\nu'),$$
$$a_m=\frac{1}{m!(\nu'-\nu+m)!}\binom{\nu'-\nu+2m}{k-\nu+m}.$$

Let $q(z)$ denote the polynomial on the left of (7), which we write as

$$q(z)=a_{m_0}z^{m_0}\left(1-\frac{a_{m_0+1}}{a_{m_0}}z\right)+a_{m_0+2}z^{m_0+2}\left(1-\frac{a_{m_0+3}}{a_{m_0+2}}z\right)+\cdots.$$

If the number of the terms is odd, the last term $a^\nu z^\nu$ will be written out separately. The important point is that this term is positive.

Since $0<\gamma<1$, it follows that $0<z\le 1/4$. On the other hand, the quotients of the coefficients can be estimated as follows:

$$\begin{aligned}\frac{a_{m+1}}{a_m}&=\frac{\nu'-\nu+2m+1}{(m+1)(\nu'-\nu+m+1)}\cdot\frac{\nu'-\nu+2m+2}{(k-\nu+m+1)(\nu'-k+m+1)}\\&<\left[\frac{1}{m+1}+\frac{1}{\nu'-\nu+m+1}\right]\cdot\left[\frac{1}{k-\nu+m+1}+\frac{1}{\nu'-k+m+1}\right]<4.\end{aligned}$$

Now it is evident that the polynomial $q(z)$ cannot have roots in the interval $(0,\,1/4]$. This contradiction proves the desired inclusion and the theorem itself.

2.3.6. THEOREM. *Let Y be a closed polydisk π-subalgebra of $C_0(^n)$ such that $Y\cap L^1(H^n)$ is C-dense in Y. If $Y\ne C_0(H^n)$, then $Y\subset B(H^n)$ or $Y\subset\overline{B}(H^n)$.*

PROOF. Suppose that Y is not contained in either of the spaces $B(H^n)$, $\overline{B}(H^n)$. By Lemmas 2.1.1, 2.1.2 and Proposition 2.1.5, it follows that the semigroup $\sigma(Y)$ defined in §2.3.1 contains real numbers with different signs. Since $\sigma(Y)$ is the closure of an open semigroup, one has $\sigma(Y)=\mathbb{R}$. Thus for almost all λ the space $E^{\lambda/2}$ is nontrivial, and there exists $\nu\in\mathbb{Z}_+^{n-1}$ such that $w_{0,\nu,\lambda/2}\in E^{\lambda/2}$ if $\lambda>0$ and $w_{0,\nu,|\lambda|/2}\in E^{\lambda/2}$ if $\lambda<0$.

Let $\lambda>0$. Setting $\alpha=\beta=w_{0,\nu,\lambda/2}$ in Lemma 2.3.4, one gets $w^2_{0,\nu,\lambda/2}\in E^\lambda$. By Lemma 2.3.5, $w_{0,k,\lambda}\in E^\lambda$ for every $k\in\mathbb{Z}_+^{n-1}$, $0\le k\le 2\nu$. In particular, $w_{0,0,\lambda}\in E^\lambda$. Similar arguments show that $w_{0,0,-\lambda}\in E^\lambda$ for $\lambda<0$.

Again, let $\lambda>0$ and $t>\lambda$. As we have just shown, $w_{0,0,t}\in E^{-t}$ and $w_{0,0,\lambda+t}\in E^{\lambda+t}$. Then by Lemma 2.3.4 the function $w_{0,0,t}\cdot w_{0,0,\lambda+t}$ belongs to the space E^λ. Choose an arbitrary multi-index $k\in\mathbb{Z}_+^{n-1}$, and consider the integral

$$J=\int_{\mathbb{C}^{n-1}}w_{0,0,t}(\zeta)w_{0,0,\lambda+t}(\zeta)\overline{w_{k,k,\lambda}(\zeta)}\,d\overline{\zeta}\,d\zeta,$$

which is equal to

$$c\prod_{j=1}^{n-1} e^{-(\lambda+2t)|\zeta|^2} L_{k_j}^{(0)}\left(2\lambda|\zeta|^2\right) d\overline{\zeta}\, d\zeta\,, \qquad c \neq 0.$$

It is clear that J does not vanish identically for $t > \lambda$; it therefore follows from Lemma 2.2.2 that $W_k^\lambda \subset E^\lambda$. So for $\lambda > 0$ one has $E^\lambda = L^2(\mathbb{C}^{n-1})$. An analogous argument shows that $E^\lambda = L^2(\mathbb{C}^{n-1})$ for $\lambda < 0$. By Proposition 2.1.5 this implies $Y = C_0(H^n)$.

2.3.7. The next theorem, together with Theorem 2.3.6, gives a complete description of the cases in which the spaces considered above are algebras.

Let Y be a closed polydisk π-subspace of $C_0(H^n)$ such that $Y \cap L^1(H^n)$ is uniformly dense in Y. As before, define

$$E^\lambda = \mathrm{cl}_{L^2}\left\{F^\lambda\left[Y \cap L^1(H^n)\right] \cap L^2(\mathbb{C}^{n-1})\right\}.$$

Define *the spectral set* $\operatorname{sp} Y$ of Y to be the closure of the set

$$\left\{(\lambda\,,\,\nu) \in (\mathbb{R}\backslash\{0\}) \times \mathbb{Z}_+^n : W_\nu^\lambda \in E^\lambda\right\},$$

or, what is the same, the closure of the set

$$\left\{(\lambda\,,\,\nu) \in (\mathbb{R}\backslash\{0\}) \times \mathbb{Z}_+^n : \nu \in \Delta^\lambda\right\},$$

where $\Delta^\lambda \subset \mathbb{Z}_+^{n-1}$ is the set of multi-indices that appear in the expansion $E^\lambda = \bigoplus_{\nu\in\Delta^\lambda} W_\nu^\lambda$.

Call a semigroup $P \subset \mathbb{R} \times \mathbb{Z}_+^{n-1}$ *conic* if the condition $(\lambda\,,\,\nu)\,,\,(\lambda'\,,\,\nu') \in \operatorname{sp} Y$ implies $(\lambda+\lambda'\,,\,k) \in \operatorname{sp} Y$ for any $k \in \mathbb{Z}_+^{n-1}$, $0 \le k \le \nu + \nu'$.

THEOREM. *A space Y is an algebra if and only if its spectral set is a conic semigroup.*

PROOF. *Necessity.* If $Y = C_0(H^n)$, then $\operatorname{sp} Y = \mathbb{R} \times \mathbb{Z}_+^{n-1}$ and there is nothing to prove. If $Y \neq C_0(H^n)$, then $Y \subset B(H^n)$ or $Y \subset \overline{B}(H^n)$ by Theorem 2.3.6. It is enough to consider the first case.

Let $W_\nu^\lambda \subset E^\lambda$ and $W_{\nu'}^{\lambda'} \subset E^{\lambda'}$; $\lambda\,,\,\lambda' > 0$. By Lemma 2.3.4, $W_\nu^\lambda \odot W_{\nu'}^{\lambda'} \subset E^{\lambda+\lambda'}$, and by Lemma 2.3.5 this is equivalent to the inclusions $W_k^{\lambda+\lambda'} \subset E^{\lambda+\lambda'}$ for all $k \in \mathbb{Z}_+^{n-1}$ such that $0 \le k \le \nu + \nu'$. It remains to recall the definition of $\operatorname{sp} Y$.

Sufficiency. Take elements φ and ψ of the space Y^0 defined in Lemma 2.3.3. For almost all $\lambda\,,\,\tau > 0$ the functions $\varphi^{\lambda-\tau}$ and ψ^τ belong to $L^2(\mathbb{C}^{n-1})$. Consider the following orthogonal expansions:

$$\varphi^{\lambda-\tau} = \sum_{\alpha\,,\,\beta\in\mathbb{Z}_+^{n-1}} a_{\alpha\,,\,\beta} w_{\alpha\,,\,\beta\,,\,\lambda-\tau}\,, \qquad \psi^\tau = \sum_{\gamma\,,\,\delta\in\mathbb{Z}_+^{n-1}} b_{\gamma\,,\,\delta} w_{\gamma\,,\,\delta\,,\,\tau}.$$

If the coefficients $a_{\alpha,\beta}$, $b_{\gamma,\delta}$ are different from zero, then by Lemma 2.2.2 $(\lambda-\tau, \beta), (\tau, \delta) \in \operatorname{sp} Y$. By Lemma 2.3.5 we have a direct sum decomposition

$$W_\beta^{\lambda-\tau} \odot W_\delta^\tau = \bigoplus_{0\le k\le\beta+\delta} W_k^\lambda,$$

and, since $\operatorname{sp} Y$ is a conic semigroup, each direct summand is a subspace of E^λ. Therefore, the product $w_{\alpha,\beta,\lambda-\tau} w_{\gamma,\delta,\tau}$, as an element of $W_\beta^{\lambda-\tau} \odot W_\delta^\tau$, belongs to E^λ. The above series, which are convergent in the $L^2(\mathbb{C}^{n-1})$ norm, may be multiplied term by term, because the basis elements and the functions $\varphi^{\lambda-\tau}$, ψ^τ are bounded. As a result one gets a series in $L^2(\mathbb{C}^{n-1})$, each term of which belongs to E^λ. Hence $\varphi^{\lambda-\tau}\psi^\tau \in E^\lambda$, and the integral

$$(\varphi\psi)^\lambda(\zeta) = \int_0^\infty \varphi^{\lambda-\tau}(\zeta)\psi^\tau(\zeta)\, d\tau$$

also defines an element of E^λ. By Lemma 2.1.2, $\varphi\psi \in Y$. Since Y^0 is an algebra, Y is also an algebra.

2.3.8. According to the following theorem, the spectral set has an even simpler structure if the algebra contains an approximate unit: the sections Δ^λ are order ideals of $\mathbb{Z}_+^{n-1}$.

THEOREM. *Let Y be a nonzero closed polydisk π-subalgebra of $C_0(H^n)$ such that $Y \cap L^1(H^n)$ is C-dense in Y. Suppose that Y contains an approximate unit. Then the projection $\sigma = \operatorname{pr}_1 \operatorname{sp} Y$ of the spectral set onto $\mathbb{R}$ is either $\mathbb{R}_+$, or $\mathbb{R}_-$ or $\mathbb{R}$ (in the latter case $Y = C_0(H^n)$ by Theorem 2.3.6). In addition, the following condition holds: if $(\lambda, \nu) \in \operatorname{sp} Y$, then $(\lambda, k) \in \operatorname{sp} Y$ for each $k \in \mathbb{Z}_+^{n-1}$, $0 \le k \le \nu$.*

PROOF. The claim concerning the set σ follows from the fact that $\operatorname{int}\sigma$ is an open semigroup of $\mathbb{R}$ and that 0 belongs to the closure of σ (Proposition 2.3.1).

The second claim follows from Theorem 2.3.7. It is enough to consider the case $\sigma = \mathbb{R}_+$; the case $\sigma = \mathbb{R}_-$ is reduced to the former by complex conjugation, and if $\sigma = \mathbb{R}$ then, by Theorem 2.3.6, Y is the algebra $C_0(H^n)$, for which the statement is trivial.

Let $(\lambda, \nu) \in \operatorname{sp} Y$, i.e., $W_\nu^\lambda \subset E^\lambda$. By Lemma 5.2.2,

$$\int_{\mathbb{C}^{n-1}} f^\lambda(\zeta)\overline{w_{\mu,\nu,\lambda}(\zeta)}\, d\bar\zeta\, d\zeta \ne 0,$$

for some $f \in Y \cap L^1(H^n)$ and $\mu \in \mathbb{Z}_+^{n-1}$. As the integral is continuous at λ, it is different from zero in a neighborhood of λ. Applying Lemma 2.2.2 once more yields $(\lambda - \varepsilon, \nu) \in \operatorname{sp} Y$ for sufficiently small $\varepsilon > 0$. Since $\sigma = \mathbb{R}_+$, one can choose ε in such a way that $(\varepsilon, \nu') \in \operatorname{sp} Y$ for some $\nu' \in \mathbb{Z}_+^{n-1}$. By

Theorem 2.3.7 the condition $(\lambda-\varepsilon, \nu)$, $(\varepsilon, \nu') \in \operatorname{sp} Y$ implies $(\lambda, k) \in \operatorname{sp} Y$ for any multi-index $k \in \mathbb{Z}_+^{n-1}$ such that $0 \le k \le \nu + \nu'$, and this proves the claim.

COROLLARY. *Under the assumptions of Theorem* 2.3.8, Y *contains one of the algebras* $A(H^n)$, $\overline{A}(H^n)$.

PROOF. If $Y = C_0(H^n)$, then there is nothing to prove. Suppose $Y \neq C_0(H^n)$. It follows from Theorem 2.3.8 that, if a section Δ^λ, $\lambda > 0$, of $\operatorname{sp} Y$ is not empty, then Δ^λ must contain a zero multi-index corresponding to the Bergmann space $\mathscr{E}^\lambda = W_0^\lambda$. In view of the assumption about the projection of $\operatorname{sp} Y$ onto $\mathbb{R}$, there are only two possibilities:

(1) $E^\lambda = \{0\}$ for $\lambda < 0$ and $\mathscr{E}^\lambda \subset E^\lambda$ for $\lambda > 0$,

(2) $E^\lambda = \{0\}$ for $\lambda > 0$ and $\overline{\mathscr{E}^\lambda} \subset E^\lambda$ for $\lambda < 0$.

The rest of the corollary follows from Proposition 2.1.5 and Lemma 2.1.2.

2.3.9. Theorems 2.3.7 and 2.3.8 can be combined for the sake of clarity. In order to formulate the statement, we introduce the following temporary terminology. Call a semigroup $P \subset \mathbb{R}_+ \times \mathbb{Z}_+^{n-1}$ *packed* if the projection of P onto $\mathbb{R}_+$ is the whole of $\mathbb{R}_+$ and if $(\lambda, \nu) \in P$ implies $(\lambda, k) \in P$ for any $k \in \mathbb{Z}_+^{n-1}$, $0 \le k \le \nu$.

THEOREM. *The class of closed, left translation-invariant, polydisk subalgebras of* $C_0(H^n)$ *that contain a dense subset of integrable functions and an approximate unit may be visualized by the following diagram:*

$$\begin{array}{ccccccccc} \{0\} & \longrightarrow & A(H^n) & \longrightarrow & \cdots & \longrightarrow & B(H^n) & \longrightarrow & C_0(H^n) \\ \| & & & & & & & & \| \\ \{0\} & \longrightarrow & \overline{A}(H^n) & \longrightarrow & \cdots & \longrightarrow & \overline{B}(H^n) & \longrightarrow & C_0(H^n) \end{array}$$

The algebras represented by dots in the upper part of the diagram are in one-to-one correspondence with the packed semigroups $P \subset \mathbb{R}_+ \times \mathbb{Z}_+^{n-1}$, *the correspondence being given by the formula* $\operatorname{sp} Y = P$. *The algebra* $A(H^n)$ *corresponds to the semigroup* $\mathbb{R}_+ \times \{0\}$, *the algebra* $B(H^n)$ *to the semigroup* $\mathbb{R}_+ \times \mathbb{Z}_+^{n-1}$. *The lower part of the diagram contains the corresponding complex conjugate algebras.*

REMARKS. (1) If $n = 1$, the group H^n degenerates into $\mathbb{R}$, the algebras $A(H^n)$ and $B(H^n)$ coincide, and the diagram contains only four algebras. This agrees with the result of de Leeuw and Mirkil [81] discussed at the beginning of the section.

(2) Theorem 2.3.9 implies, in particular, that every packed semigroup corresponds to an invariant algebra from the class under consideration. To save

space, we do not prove this statement. However, the construction of an invariant algebra, given its spectral set, and the verification of its properties offer no difficulty.

2.3.10. An interesting class of invariant algebras is connected with superlinear maps.

Let $\varphi : \mathbb{R}_+ \to \mathbb{Z}_+^{n-1} \cup \infty$ be a superlinear map. Define Y_φ as the uniform closure of the set Z of functions $f \in L^1(H^n) \cap C_0^\infty(H^n)$ such that

$$D_{\bar\zeta}^{\nu}\left(e^{\lambda|\zeta|^2} f^\lambda\right) = 0, \qquad \nu \notin [0, \varphi(\lambda)], \ \lambda > 0; \quad f^\lambda = 0, \quad \lambda < 0. \tag{1}$$

It is clear that the space $E^\lambda = \mathrm{cl}_{L^2}\left[f^\lambda(Z) \cap L^2(\mathbb{C}^{n-1})\right]$, $\lambda > 0$, contains the Bergmann space $\mathscr{E}^\lambda$; therefore, by Proposition 2.1.5, $A(H^n) \subset Y_\varphi \subset B(H^n)$. Note that the algebras $A(H^n)$, $B(H^n)$ correspond to the maps $\varphi(\lambda) \equiv 0$, $\varphi(\lambda) \equiv \infty$, respectively.

The space Y_φ is a polydisk π-invariant space. Moreover, it is an algebra. Indeed, let f and g satisfy (1). Then the derivatives

$$D_{\bar\zeta}^{\nu}\left(e^{\lambda|\zeta|^2}(fg)^\lambda\right) = \sum_{0 \le k \le \nu} \binom{\nu}{k} \int_{\mathbb{R}} D_{\bar\zeta}^{k}\left(e^{-(\lambda-\tau)|\zeta|^2} f^{\lambda-\tau}\right) D_{\bar\zeta}^{\nu-k}\left(e^{\tau|\zeta|^2} g^\tau\right) d\tau$$

vanish for $\nu \notin [0, \varphi(\lambda)]$, $\lambda > 0$, because the inequalities $k \le \varphi(\lambda - \tau)$, $\nu - k \le \varphi(\tau)$ cannot hold simultaneously for a superlinear map, so that all the integrands vanish. The second condition in (1) is obviously true.

The spectral set of Y_φ is subgraph of the map φ:

$$\mathrm{sp}\, Y_\varphi = \left\{(\lambda, \nu) \in \mathbb{R}_+ \times \mathbb{Z}_+^{n-1} : 0 \le \nu \le \varphi(\lambda)\right\}.$$

Indeed, by formula (4), §2.2.1, the functions $e^{\lambda|\zeta|^2} w_{\mu,\nu,\lambda}$ are polynomials of degree ν in $\bar\zeta$; hence, the Fourier expansions of the functions f^λ, $f \in Z$, $\lambda > 0$, contain only terms $w_{\mu,\nu,\lambda}$ with $\nu \le \varphi(\lambda)$.

Note that in the case $n = 2$ the class of closed polydisk π-subalgebras of $C_0(H^n)$ with a dense subset of integrable functions and an approximate unit is exhausted by the algebras of type Y_φ or $\overline{Y}_\varphi$. This follows from Theorem 5.3.8, because all the order ideals Δ^λ of $\mathbb{Z}_+$ have the form $\Delta^\lambda = [0, \varphi(\lambda)]$, $\varphi(\lambda) = \sup \Delta^\lambda$, where φ is a superlinear map, according to Theorem 5.3.7.

2.4. Anisotropic π-subalgebras of $C_0(H^n)$. According to the following lemma, if an invariant space is anisotropic, it automatically contains a dense subset of integrable functions.

2.4.1. Lemma. *Let Y be a closed anisotropic π-subspace of $C_0(H^n)$. Then $Y \cap L^1(H^n)$ is dense in Y.*

PROOF. Let N denote the closure in $L^2(H^n)$ of the set of all $h \in L^1(H^n) \cap L^2(H^n)$ such that

$$\int_{H^n} f\bar{h}\,dx\,d\bar{\zeta}\,d\zeta = 0$$

for any $f \in Y$.

Then N is an anisotropic π-subspace of $L^2(H^n)$, and for almost all λ one has an orthogonal decomposition

$$N^\lambda = \mathrm{cl}_{L^2}\left[F^\lambda(Z) \cap L^2(\mathbb{C}^{n-1})\right] = \bigoplus_{\nu \in \Delta^\lambda} W_\nu^\lambda,$$

where, by Lemma 2.2.4, the sets Δ^λ are the same for all λ of the same sign. It is therefore natural to set $\Delta^\lambda = \Delta^\pm$ for $\lambda \in \mathbb{R}_\pm$.

Define Y_1 as the set of all $f \in L^1(H^n) \cap C_0(H^n)$ such that

$$\begin{aligned} f(x, \zeta) = \frac{1}{2\pi} \int_0^\infty e^{i\lambda x} \sum_{\mu, \nu \in \mathbb{Z}_+^{n-1}} a_{\mu,\nu}^+(\lambda) w_{\mu,\nu,\lambda}(\zeta)\,d\lambda \\ + \frac{1}{2\pi} \int_0^\infty e^{i\lambda x} \sum_{\mu, \nu \in \mathbb{Z}_+^{n-1}} a_{\mu,\nu}^-(\lambda) w_{\nu.\mu,|\lambda|}(\zeta)\,d\lambda, \end{aligned}$$

where $a_{\mu,\nu}^\pm(\lambda) = 0$ for $\nu \in \Delta^\pm$ and $\lambda \in \mathbb{R}_\pm$. We will prove that Y_1 is uniformly dense in Y.

For any $h \in L^1(H^n) \cap L^2(H^n)$,

$$\int_{H^n} f\bar{h}\,dx\,d\bar{\zeta}\,d\zeta = \sum_{\mu,\nu \in \mathbb{Z}_+^{n-1}} \int_0^\infty a_{\mu,\nu}^+(\lambda) h_{\mu,\nu}^+(\lambda)\,d\lambda + \sum_{\mu,\nu \in \mathbb{Z}_+^{n-1}} \int_0^\infty a_{\mu,\nu}^-(\lambda) \overline{h_{\mu,\nu}^-(\lambda)}\,d\lambda, \tag{1}$$

where

$$\begin{aligned} h_{\mu,\nu}^+(\lambda) &= \int_{\mathbb{C}^{n-1}} h^\lambda(\zeta) \overline{w_{\mu,\nu,\lambda}(\zeta)}\,d\bar{\zeta}\,d\zeta, \qquad \lambda > 0, \\ h_{\mu,\nu}^-(\lambda) &= \int_{\mathbb{C}^{n-1}} h^\lambda(\zeta) \overline{w_{\nu,\mu,|\lambda|}(\zeta)}\,d\bar{\zeta}\,d\zeta, \qquad \lambda < 0. \end{aligned}$$

It follows from (1) that every function $h \in N$ is orthogonal to Y_1, because of the special choice of the functions $a_{\mu,\nu}^\pm$. Conversely, if h is orthogonal to all $f \in Y_1$, then (1) implies that $h_{\mu,\nu}^\pm = 0$ for $\nu \notin \Delta^\pm$, since the choice of $a_{\mu,\nu}^\pm$ is sufficiently flexible. Therefore, $h^\lambda \in N^\lambda$ for almost all λ. By Lemma 2.1.2, $h \in N$.

Thus, Y and Y_1 have the same annihilators in $L^1(H^n) \cap L^2(H^n)$. Since absolutely continuous measures of the form $\bar{h}dxd\bar{\zeta}d\zeta$, $h \in L^1(H^n) \cap L^2(H^n)$, form a weakly dense subset of the space of measures that annihilate a given closed invariant subspace of $C_0(H^n)$ (see Lemma 2.1.1), it follows that $\mathrm{cl}_C Y_1 = Y$. This completes the proof of the theorem.

2.4.2. THEOREM. *Let Y be a closed anisotropic polydisk π-subalgebra of the space $C_0(H^n)$. Suppose that Y is invariant with respect to all permutations σ of the variables $\zeta_1, \dots, \zeta_{n-1}$, i.e., $f_\sigma(x, \zeta) = f(x, \sigma\zeta) \in Y$ for any $f \in Y$. Then Y is one of the algebras in the following diagram:*

$$\begin{array}{ccccccc} \{0\} & \longrightarrow & A(H^n) & \longrightarrow & B(H^n) & \longrightarrow & C_0(H^n) \\ \| & & & & & & \| \\ \{0\} & \longrightarrow & \overline{A}(H^n) & \longrightarrow & \overline{B}(H^n) & \longrightarrow & C_0(H^n) \end{array}$$

PROOF. Suppose $Y \neq C_0(H^n)$. Then, by Lemma 2.4.1 and Theorem 2.3.6, either $Y \subset B(H^n)$ or $Y \subset \overline{B}(H^n)$. Suppose $Y \subset B(H^n)$ and $Y \neq \{0\}$. By Lemma 2.2.4, since Y is anisotropic, the spectral set of Y is $\operatorname{sp} Y = \mathbb{R}_+ \times \Delta$, $\Delta \subset \mathbb{Z}_+^{n-1}$.

Suppose that $\Delta = \{0\}$. By the definition of $\operatorname{sp} Y$ and Proposition 2.1.5, this means that $Y = A(H^n)$.

Now suppose that $\Delta \neq \{0\}$, and let $\nu \in \Delta$ be a multi-index such that $\nu_j > 0$. Then, by Theorem 2.3.7, Δ contains all $k \in \mathbb{Z}_+^{n-1}$, $0 \le k \le 2\nu$. In particular, Δ contains the multi-index $\delta_j = (0, \dots, 1, \dots, 0)$. Permuting the variables ζ_j, we conclude that Δ contains all the multi-indices δ_j, $j = 1, \dots, n-1$. Since Δ is a semigroup (Theorem 2.3.7), this implies that $\Delta = \mathbb{Z}_+^{n-1}$, but then by Proposition 2.3.7 $Y = B(H^n)$.

The case $Y \subset \overline{B}(H^n)$ is reduced to the previous one by complex conjugation.

2.5. Möbius function algebras on a sphere. The Cayley transformation maps the group of affine analytic automorphisms of the Siegel domain into the subgroup $\mathrm{Aut}(B^n, e)$ of analytic automorphisms of the ball B^n that leave the point $e = (1, 0, \dots, 0)$ on the sphere S fixed. The algebra $C_0(H^n)$ corresponds to the algebra $C_0(S)$ of continuous functions on S that vanish at e. The algebra $A(H^n)$ is transformed into the algebra $A_0(S) = A(S) \cap C_0(S)$, where $A(S)$ is the set of the boundary values of functions holomorphic in the open unit ball and continuous in the closed unit ball. The algebra $B(H^n)$ is transformed into the algebra $B_0(S) = B(S) \cap C_0(S)$, where $B(S)$ is the set of those elements of $C(S)$ which may be continued continuously to the closed unit ball, and the continuation is analytic on every complex straight line through e that intersects the open ball.

2.5.1. The following theorem is a direct consequence of Theorem 2.4.2.

THEOREM. *A closed subalgebra of $C_0(S)$ that is invariant under the action of the group $\mathrm{Aut}(B^n, e)$ is one of the entries in the following diagram:*

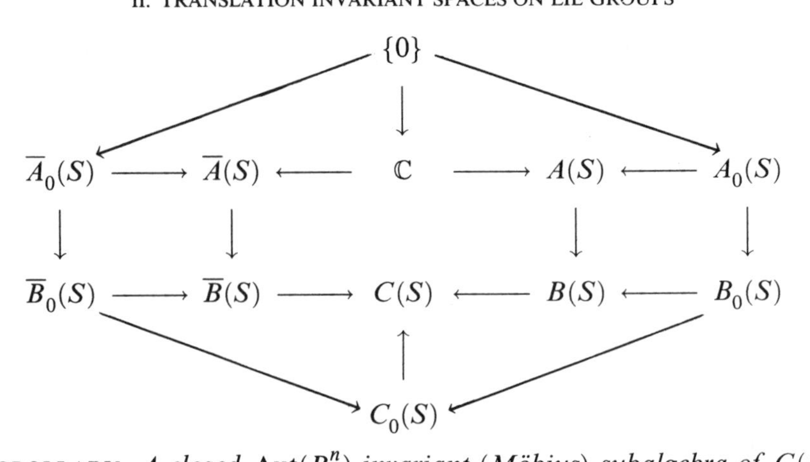

Corollary. *A closed* $\mathrm{Aut}(B^n)$*-invariant (Möbius) subalgebra of* $C(S)$ *is one of the entries in the following diagram*:

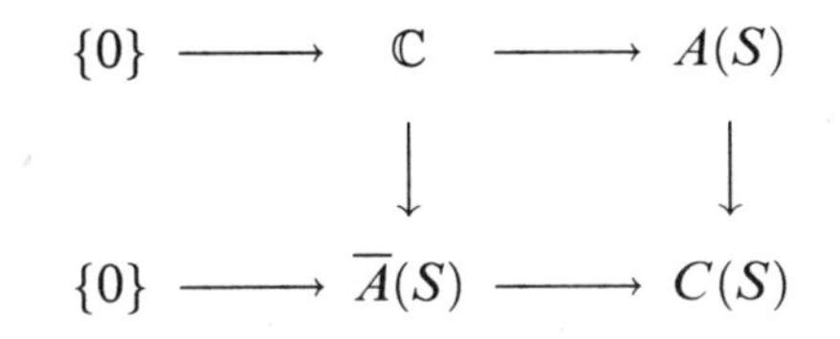

2.6. π**-subspaces of** $C_0(H^n)$ **invariant under infinitesimal nonlinear automorphisms. Möbius function spaces on the sphere.**

2.6.1. The automorphisms of the ball B^n for which e is not a fixed point are mapped by the Cayley transformation into nonlinear automorphisms of the Siegel domain Ω^n that map the point at infinity onto a finite point.

Consider the one-parameter group of such automorphisms corresponding to the group of rotations of the ball with respect to the variable w_1:

$$T_t(w_1, \dots, w_n) = \left(e^{it}w_1, w_2, \dots, w_n\right), \qquad t \in \mathbb{R}.$$

The corresponding transformation group of H^n consists of the maps

$$T_t^*(x, \zeta) = \left(-\operatorname{Im}\frac{z(1+e^{it}) + i(1-e^{it})}{z(1-e^{it}) + i(1+e^{it})}, i\zeta_1, \dots, i\zeta_{n-1}\right), \qquad z = x + i|\zeta|^2.$$

Consider the generator of this group:

$$\begin{aligned} D_0 f &= \frac{d}{dt}\left(f \circ T_t^*\right)\Big|_{t=0} \\ &= \frac{1}{2}\operatorname{Re}(1+z^2)\frac{\partial f}{\partial x} + \frac{1}{2}\sum_{k=1}^{n-1}(z-i)\zeta_k\frac{\partial f}{\partial \zeta_k} + \frac{1}{2}\sum_{k=1}^{n-1}(\overline{z}+i)\overline{\zeta}_k\frac{\partial f}{\partial \overline{\zeta}_k}. \quad (1) \end{aligned}$$

To simplify matters, let us consider the operator

$$D = \operatorname{Re} z^2\frac{\partial}{\partial x} + z\sum_{k=1}^{n-1}\zeta_k\frac{\partial}{\partial \zeta_k} + \overline{z}\sum_{k=1}^{n-1}\overline{\zeta}_k\frac{\partial}{\partial \overline{\zeta}_k}, \qquad (2)$$

which is obtained by subtracting from $2D_0$ the generator $\partial/\partial x$ of the group of translations in x and the generator $i\sum_{k=1}^{n-1}(\overline{\zeta}_k(\partial/\partial\overline{\zeta}_k)-\zeta_k(\partial/\partial\zeta_k)$ of the rotation group $(x, \zeta)\to(x, e^{-it}\zeta)$ in the variable ζ. The Fourier transform of the differential operator in the variable x is

$$\begin{aligned}\frac{1}{i}\widehat{D} &= -\left(\frac{d^2}{d\lambda^2}+|\zeta|^4\right)\lambda+\sum_{k=1}^{n-1}\frac{\partial}{\partial\lambda}\left(\zeta_k\frac{\partial}{\partial\zeta_k}+\overline{\zeta}_k\frac{\partial}{\partial\overline{\zeta}_k}\right)\\ &\quad+|\zeta|^2\sum_{k=1}^{n-1}\left(\zeta_k\frac{\partial}{\partial\zeta_k}-\overline{\zeta}\frac{\partial}{\partial\overline{\zeta}_k}\right).\end{aligned}\tag{3}$$

Our primary goal is to examine the action of operator $\widehat{D}$ on the basis elements $w_{\mu,\nu,\lambda}$.

Take an arbitrary element $\nu\in\mathbb{Z}_+^{n-1}$, $\nu\neq 0$, and an arbitrary $\lambda>0$. It is convenient to use the functions

$$w'_{0,\nu,\lambda}(\zeta)=e^{-\lambda|\zeta|^2}\overline{\zeta}^{\nu},$$

which, by (4), §2.2.1, differ from $w_{0,\nu,\lambda}$ only by a positive factor depending on λ (and of course also on ν).

Calculation yields

$$\frac{1}{i}\widehat{D}w'_{0,\nu,\lambda}=-2|\nu|\,|\zeta|^2e^{-\lambda|\zeta|^2}\sum_{k=1}^{n-1}\zeta_k\overline{\zeta}^{\nu+\delta_k},$$

where $\delta_k=(0,\dots,1,\dots,0)$, with the unit occupying the kth place.

Each of the summands is easily expressed in terms of the basis elements by (4), §2.2.1, and by the relations

$$w_{0,\nu,\lambda}(\zeta)=c_{0,\nu,\lambda}e^{-\lambda|\zeta|^2}\left(\sqrt{2\lambda}\overline{\zeta}\right)^{\nu},$$

$$w_{\delta_k,\nu+\delta_k,\lambda}(\zeta)=c_{\delta_k,\nu+\delta_k,\lambda}e^{-\lambda|\zeta|^2}\left[\left(\sqrt{2\lambda}\overline{\zeta}\right)^{\nu+\delta_k}\left(\sqrt{2\lambda}\zeta_k\right)-(\nu_k+1)\left(\sqrt{2\lambda}\overline{\zeta}\right)^{\nu}\right].$$

Finally we see that $\widehat{D}w'_{0,\nu,\lambda}$ can be represented as

$$\frac{1}{i}\widehat{D}w'_{0,\nu,\lambda}=a_0(\lambda)w_{0,\nu,\lambda}(\zeta)+\sum_{k=1}^{n-1}a_k(\lambda)w_{\delta_k,\nu+\delta_k,\lambda}(\zeta),$$

where the coefficients $a_k(\lambda)$, $k=0, 1, \dots,$ do not vanish.

Thus, applying the operator $\widehat{D}$ to a function $w'_{0,'\nu,\lambda}$, we may enlarge the component of ν by one.

The action of $\widehat{D}$ on the other basis elements decreases ν. Namely, consider the function

$$w'_{\delta_k,\nu,\lambda}(\zeta)=e^{-\lambda|\zeta|^2}\left(-2\lambda\overline{\zeta}^{\nu}\zeta_k+\nu_k\overline{\zeta}^{\nu-\delta_k}\right),\qquad \nu_k>0,$$

which differs from $w_{\delta_k,\nu,\lambda}$ by a factor depending on λ. Direct calculation yields

$$\frac{1}{i}\widehat{D}w_{\delta_k,\nu,\lambda}(\zeta)=2e^{-\lambda|\zeta|^2}\left[(1-|\nu|)\overline{\zeta}^{\nu}\zeta_k+2\lambda|\nu|\,|\zeta|^2\overline{\zeta}^{\nu}\zeta_k-\nu_k(|\nu|-1)|\zeta|^2\overline{\zeta}^{\nu-\delta_k}\right].$$

A straightforward calculation proves that the scalar product

$$\int_{\mathbb{C}^{n-1}}\left(\widehat{D}w'_{\delta_k,\nu,\lambda}\right)(\zeta)\overline{w_{0,\nu-\delta_k,\lambda}(\zeta)}\,d\overline{\zeta}\,d\zeta$$

is positive, and therefore the orthogonal expansion of $\widehat{D}w'_{\delta_k,\nu,\lambda}$ in terms of the λ-basis contains the term $w_{0,\nu-\delta_k,\lambda}$.

This argument proves the following

Lemma. *Let $\nu\in\mathbb{Z}_+^{n-1}$, $\nu\neq 0$. There exist smooth positive factors $\rho(\lambda)$, $\sigma(\lambda)$ such that the expansion of the function $\widehat{D}(\rho w_{0,\nu,\lambda})$ in terms of the λ-basis contains all the elements $w_{\delta_k,\nu+\delta_k,\lambda}$, $k=1,\dots,n-1$, and that of the function $\widehat{D}(\sigma w_{\delta_k,\nu,\lambda})$ contains all the elements $w_{0,\nu-\delta_k,\lambda}$ with k such that $\nu_k>0$.*

2.6.2. Let Y be a closed π-subspace of $C_0(H^n)$. Since the convolution of an element of Y by a smooth function on H^n with compact support again belongs to Y, it follows that $Y\cap C^\infty(H^n)$ is uniformly dense in Y. Define

$$Y_D=\left\{f\in Y\cap C^1(H^n):\ Df\in C_0(H^n)\right\}.$$

Theorem. *Let Y be a closed anisotropic polydisk π-subspace of $C_0(H^n)$. Suppose that the operator D maps Y_D onto Y. Then Y is with one of the entries in the following diagram:*

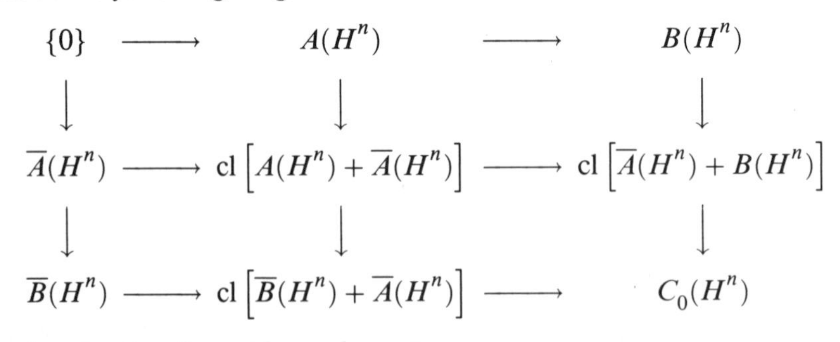

($\mathrm{cl}=\mathrm{cl}_C$ *denotes the uniform closure*).

Proof. By Lemma 2.4.1, $Y\cap L^1(H^n)$ is uniformly dense in Y. Since Y is anisotropic, it follows from Lemma 2.3.3 that the spaces

$$E^\lambda=\mathrm{cl}_{L^2}\left\{F^\lambda[Y\cap L^1(H^n)]\cap L^2(\mathbb{C}^{n-1})\right\}$$

are identical for λ of the same sign. Let $\lambda_0 > 0$, and suppose $E^{\lambda_0} \neq \{0\}$. If E^{λ_0} is not the Bergmann space $\mathscr{E}^{\lambda_0} = W_0^{\lambda_0}$, then $w_{0,\nu,\lambda_0} \in E^{\lambda_0}$ for some $\nu \neq 0$.

Let $\gamma(\lambda) \geq 0$ be a smooth function with compact support such that $\gamma(\lambda)$ coincides with the function $\rho(\lambda)$ defined in Lemma 2.6.1 in a sufficiently small neighborhood V of λ_0. Since $w_{0,\nu,\lambda} \in E^{\lambda}$ for almost all positive λ, it follows from Lemma 2.1.2 that the function

$$f(x, \zeta) = \int_0^\infty e^{i\lambda x} \gamma(\lambda) w_{0,\nu,\lambda}(\zeta)\, d\lambda$$

belongs to $\mathrm{cl}_{L^2}\left[Y \cap L^1(H^n)\right]$. Since $f \in C_0(H^n)$, it follows from Lemma 2.1.1 that $f \in Y$.

By assumption, $Df \in Y$, because $Df \in C_0(H^n)$. Then $(Df)^\lambda = \widehat{D} f^\lambda \in E^\lambda$ for positive λ. But $f^\lambda = \gamma(\lambda) w_{0,\nu,\lambda}$; therefore, by Lemma 2.6.1, the spaces E^λ, $\lambda \in V$, contain elements $w_{\delta_k, \delta_k+\nu, \lambda}$. Hence $W^\lambda_{\nu+\delta_k} \subset E^\lambda$. In the same way, Lemma 2.6.1 implies $W^\lambda_{\nu-\delta_k} \subset E^\lambda$ if $\nu_k > 0$. It follows that, if $E^{\lambda_0} \neq \mathscr{E}^{\lambda_0}$, then $E^{\lambda_0} = L^2(\mathbb{C}^{n-1})$.

Thus, only three situations are possible for all positive λ_0:

$$E^{\lambda_0} = \{0\}, \quad E^{\lambda_0} = \mathscr{E}^{\lambda_0}, \qquad E^{\lambda_0} = L^2(\mathbb{C}^{n-1}).$$

In case $\lambda_0 < 0$ the possibilities are

$$E^{\lambda_0} = \{0\}, \qquad E^{\lambda_0} = \overline{\mathscr{E}^{\lambda_0}}, \qquad E^{\lambda_0} = L^2(\mathbb{C}^{n-1}).$$

The proof is completed by applying Lemmas 2.1.2, 2.1.3, 2.1.4 and Proposition 2.1.5.

2.6.3. The Cayley transformation carries the operator D_0 on the Heisenberg group to the differential operator

$$df(w) = \frac{d}{dt} f\left(e^{it} w_1, w_2, \dots, w_n\right)\Big|_{t=0}$$

on the unit sphere $S \subset \mathbb{C}^{n-1}$.

THEOREM. *A closed subspace of* $C_0(H^n)$ *which is invariant under the action of* $\mathrm{Aut}(B^n, e)$ *and in which the differential operator* d *acts in the same sense as in Theorem* 2.6.2 *is one of the entries in the following diagram*:

$$\begin{array}{ccccc}
\{0\} & \longrightarrow & A_0(S) & \longrightarrow & B_0(S) \\
\downarrow & & \downarrow & & \downarrow \\
\overline{A}_0(S) & \longrightarrow & \mathrm{Plh}_0(S) & \longrightarrow & \mathrm{cl}\left[\overline{A}_0(S) + B_0(S)\right] \\
\downarrow & & \downarrow & & \downarrow \\
\overline{B}_0(S) & \longrightarrow & \mathrm{cl}\left[\overline{B}_0(S) + A_0(S)\right] & \longrightarrow & C_0(S)
\end{array}$$

A list of the subspaces of $C(S)$ *invariant under* $\mathrm{Aut}(B^n, e)$ *is obtained by omitting the subscript 0 in the spaces listed in this diagram and adding the space* $\mathbb{C}$.

Here $\mathrm{Plh}(S)$ *is the space of functions that have a continuous pluriharmonic continuation to the ball* B^n; *the subscript* 0 *means that only functions that vanish at* e *are included.*

This theorem is obtained by carrying Theorem 2.6.2 over from H^n to S. We need only note that D_0-invariance of spaces on H^n, together with invariance under the transformations $(x, \zeta) \to (x + a, e^{it}\zeta)$, $a, t \in \mathbb{R}$, is equivalent to d-invariance of the corresponding spaces on the sphere, by the definition of d and D_0.

2.6.4. Any additional symmetry condition naturally decreases the number of invariant spaces.

In particular, the spaces $B(S)$, $\mathrm{cl}_C[\overline{A}(S) + B(S)]$, and their complex conjugates are not invariant under reflection of the sphere in the complex hyperplane $w_1 = 0$:

$$(w_1, \dots, w_n) \to (-w_1, w_2, \dots, w_n).$$

Hence the condition of invariance under the full group of analytic automorphism of the ball excludes these spaces from the list of Theorem 2.6.2, yielding the theorem of Nagel and Rudin [84]: A closed Möbius subspace of $C(S)$ is one of the following:

$$\{0\}, \ \mathbb{C}, \ A(S), \ \overline{A}(S), \ \mathrm{Plh}(S), \ C(S).$$

This was proved in [84] by harmonic analysis on the unitary group.

CHAPTER III

Möbius Spaces and Algebras on Symmetric Domains and Their Shilov Boundaries

From this chapter on we focus our attention on invariant spaces and algebras in Hermitian symmetric domains of noncompact type. In this case isometries are just holomorphic diffeomorphisms. According to E. Cartan's theory, the Hermitian symmetric spaces of noncompact type are precisely the bounded symmetric domains in a space of several complex variables. This realization yields two natural homogeneous spaces of the isometry group—the domain itself and its Shilov boundary.

The main topics of this chapter are function spaces and algebras on these two homogeneous spaces that are invariant under analytic automorphisms. We will show that invariant spaces and algebras are formed in one way or another from holomorphic functions or their boundary values. This yields new characterizations of holomorphic functions and their boundary values for symmetric domains.

§1. Affine-invariant function algebras on Shilov boundaries of Siegel domains of the first kind

Let D be a bounded symmetric domain in $\mathbb{C}^n$. Recall that this means that every point of D is an isolated fixed point of an involutive analytic automorphism of D. Every bounded symmetric domain is isomorphic to an unbounded Siegel domain of the first or second kind [33], i.e., either to a tube $\Omega_V = \mathbb{R}^n + iV$ over an open homogeneous convex pointed cone, or to a domain

$$\Omega_{V,F} = \left\{(z, \zeta) \in \mathbb{C}^n \times \mathbb{C}^m : \operatorname{Im} z - F(\zeta, \zeta) \in V\right\},$$

where V is a cone with the same properties as above, $F : \mathbb{C}^m \times \mathbb{C}^m \to \mathbb{R}^n$ a nondegenerate semihermitian form. Symmetric domains correspond to selfdual cones.

Throughout this section $\Omega = \Omega_V$ is a tube over a cone V. Let $\mathscr{A}(V)$ denote the group of affine maps of V onto itself and $\mathscr{A}(\Omega)$ the group of affine analytic automorphisms of Ω of the type

$$z \to a + ux + iuy, \qquad z = x + iy,$$

where $a \in \mathbb{R}^n$, $u \in \mathscr{A}(V)$.

Let V be a convex (maybe flat) cone in $\mathbb{R}^n$. Denote the complexification of the linear space l_V spanned by V by $l_V^{\mathbb{C}}$. Set

$$\Omega_V = \left\{ z \in l_V^{\mathbb{C}} : \operatorname{Im} z \in V \right\}.$$

By the Shilov boundary $\partial\Omega_V$ of Ω_V we mean the space l_V.

Let $\sigma(f)$, $f \in L^\infty(\mathbb{R}^n)$, denote the Beurling spectrum of f, i.e., the set of common zeros of the Fourier transforms of functions $h \in L^1(\mathbb{R}^n)$ such that $h * f = 0$.

1.1. LEMMA. *Let $f \in L^\infty(\mathbb{R}^n) \cap C(\mathbb{R}^n)$. Suppose that for any $a \in \mathbb{R}^n$ the restriction $f|_{a+\partial\Omega_V}$ can be extended to a function holomorphic and bounded in $a + \Omega_V$. Then $\sigma(f)$ is contained in the closure $\overline{V}^*$ of the dual cone V^*.*

PROOF. After a suitable orthogonal transformation of $\mathbb{R}^n$, we may assume that V is contained in the octant

$$\mathbb{R}^n_+ = \left\{ y \in \mathbb{R}^n : y_1 \geq 0, \dots, y_m \geq 0, y_{m+1} = \dots = y_n = 0 \right\},$$

where m is the dimension of V.

Let π_1 denote orthogonal projection $(y_1, \dots, y_m, y_{m+1}, \dots, y_n) \to (y_1, \dots, y_m)$, and π_2 orthogonal projection $(y_1, \dots, y_m, y_{m+1}, \dots, y_n) \to (y_{m+1}, \dots, y_n)$. By assumption, for any $a_2 \in \mathbb{R}^{n-m}$ the restriction $f_{a_2} = f|_{\pi_2^{-1}(a_2)}$, as a function of $a_1 \in \mathbb{R}^m$, may be continued holomorphically into the m-dimensional complex region Ω_V. Let us prove that $\sigma(f_{a_2}) \subset \overline{V}^*$. It is enough to do this for $a_2 = 0$. Consider the integrable factors

$$\alpha_k(z) = \frac{(-1)^m k^{2m}}{\prod_{j=1}^{m} (z_j + ik)^2}.$$

Since the products $\alpha_k f_0$ belong to the Hardy class $H^2(\partial\Omega_V)$, the following integral representation is valid [18], [29]:

$$(\alpha_k f_0)(x) = \int_{V^* \cap \mathbb{R}^m} e^{i\langle x, t\rangle} g_k(t)\, dt, \qquad x \in \mathbb{R}^m,$$

where $g_k \in L^2(V^* \cap \mathbb{R}^m)$.

This implies that $\sigma(\alpha_k f_0) \subset \overline{V}^* \cap \mathbb{R}^m$. The sequence α_k is bounded and converges pointwise to 1; thus letting $k \to \infty$ we obtain $\sigma(f_0) \subset \overline{V}^* \cap \mathbb{R}^m$. The same argument shows that for any $a_2 \in \mathbb{R}^{n-m}$ we have $\sigma(f_{a_2}) \subset \overline{V}^* \cap \mathbb{R}^m$. This yields

$$\sigma(f) \subset \left(\overline{V}^* \cap \mathbb{R}^m \right) \oplus \mathbb{R}^{n-m} = \overline{V}^*.$$

1.2. For an open convex cone V in $\mathbb{R}^n$ let $A(\partial\Omega_V)$ denote the algebra of all continuous functions in $\mathbb{R}^n = \partial\Omega_V$ that can be extended to a function holomorphic in Ω_V and continuous in $\mathbb{R}^n \cup \Omega_V \cup \{\infty\}$. If V is a flat cone, then $A(\partial\Omega_V)$ will denote the algebra of all functions $f \in C(\mathbb{R}^n)$ that admit an extension to a function continuous in $\mathbb{R}^n \cup (\mathbb{R}^n + iV) \cup \{\infty\}$ and holomorphic in every fiber $a + \partial\Omega_V$.

LEMMA. *Let V be a convex pointed cone in $\mathbb{R}^n$ and f an element of $C_0(\mathbb{R}^n)$ such that $\sigma(f) \in V^*$. Then $f \in A(\partial\Omega_V)$.*

PROOF. Since V is a pointed cone, the dual cone coincides with the closure of its interior. Hence we can approximate f uniformly by a sequence $f_k \in L^1(\mathbb{R}^n)$ of functions with compact support (see, for example, [81]). We can choose f_k with the additional property that their Fourier transforms are C^∞ smooth.

Continue f_k to an entire function $\widetilde{f_k}$ of n variables, defined by

$$\widetilde{f_k}(z) = \frac{1}{(2\pi)^n} \int_{\sigma(f_k)} e^{i\langle z, t\rangle} \hat{f}_k(t)\, dt.$$

The formula for the derivatives of Fourier transforms and the compactness of the spectrum $\sigma(f_k) \subset V^*$ imply the estimate:

$$\left|\widetilde{f_k}(z)\right| \le C(1 + |z|^2)^{-1} \cdot \sup_{t \in \sigma(f_k)} e^{-\langle \operatorname{Im} z, t\rangle}.$$

If $\operatorname{Im} z \in V$, then

$$\inf\{\langle \operatorname{Im} z, t\rangle : t \in \sigma(f_k), \operatorname{Im} z \in V\} > 0,$$

hence, $\lim_{z\to\infty, z\in\Omega_V} \widetilde{f_k}(z) = 0$. By the maximum principle, the functions $f_k|_{a+\partial\Omega_V}$ converge uniformly as $k \to \infty$ to a function f_a which is analytic in $a + \Omega_V$ and vanishes at infinity. It is clear that f and f_a have the same restrictions to $a + \partial\Omega_V$.

1.3. THEOREM. *Let A be a closed, dilatation- and translation-invariant subalgebra of $C(\overline{\mathbb{R}}^n)$. Then the Beurling spectrum $W = \sigma(A) = \overline{\cup_{f\in A}\sigma(f)}$ is a convex cone, and $A = A(\partial\Omega_V)$, where $V = \operatorname{int}_{l(W^*)} W^*$.*

PROOF. Since A is an algebra, it follows that $\sigma(A)$ is a semigroup (see, e.g., [81]). In addition, it follows from the assumptions that $\sigma(A)$ is dilatation-invariant, so $\sigma(A)$ is a convex cone.

If A does not contain nonconstant functions, then $\sigma(A) = \{0\}$, and $[\sigma(A)]^* = \mathbb{R}^n$. Then $A(\partial\Omega_V)$ consists of bounded entire functions, which are constants by the Liouville theorem. In this case, therefore, the proof is complete. Now let $A \ne \mathbb{C}$. Then $A \cap C_0(\mathbb{R}^n) \ne \{0\}$; therefore, $\operatorname{int}\sigma(A) \ne \varnothing$ and the convex cone $\sigma(A)$ coincides with the closure of its interior. The result now follows from Lemmas 1.1 and 1.2.

1.4. THEOREM. *Let V be an affine-homogeneous cone. Then every closed $\mathscr{A}(\Omega_V)$-invariant subalgebra A of $C(\mathbb{R}^n)$ that contains $\mathscr{A}(\partial\Omega_V)$ as a proper subset has the form $A = A(\partial\Omega_\gamma)$, where γ is a face of the closed cone $\overline{V}$.*

PROOF. It follows from Theorem 1.3 that $A = A(\partial\Omega_\gamma)$, where γ is the interior of the cone $[\sigma(A)]^*$ with respect to its linear span. Since $A(\partial\Omega_\gamma) \subset A$, it follows that $V^* = \sigma\left(A(\partial\Omega_V)\right) \subset \sigma(A)$. The dual cones satisfy the opposite inclusion $\overline{\gamma} \subset (V^*)^* = \overline{V}$. Since A is invariant, γ is invariant under affine automorphisms of $\overline{V}$. Since V is affine-homogeneous, either $\gamma = V$, which implies $A = A(\partial\Omega_V)$, or γ is contained in the boundary of the closed cone $\overline{V}$ and is a face of the latter.

The sets $\Omega_\gamma = \{x + iy \in l_\gamma^{\mathbb{C}} : \operatorname{Im} y \in \gamma\}$ are called the linear components of the boundary of the Siegel domain Ω_V. Thus, Theorem 1.4 states that any closed $\mathscr{A}(\Omega_V)$-invariant subalgebra of $C(\mathbb{R}^n)$ that contains the algebra of boundary values (on the Shilov boundary) of holomorphic functions may be defined by the condition that there exists a holomorphic extension to the linear components of the boundary.

§2. Boundary values of holomorphic functions on the Shilov boundary of bounded symmetric domain

In this section we establish conditions for the existence of holomorphic continuations of continuous functions from the Shilov boundary of a bounded symmetric domain to the domain itself. The results of this section will be used in §3 to describe invariant algebras on the Shilov boundary.

Let D be a bounded symmetric domain in $\mathbb{C}^n$. Choose a base point $o \in D$, and let K denote the stationary subgroup of the group $\operatorname{Aut}(D)$ of all analytic automorphisms of D. Let σ denote the (unique) normalized K-invariant measure on the Shilov boundary πD and π the canonical projection $\operatorname{Aut}(D) \to D$; $\pi(\omega) = \omega(o)$.

Let P denote the Poisson operator for the Laplace-Beltrami operator in D. This operator extends functions $f \in C(\partial D)$ to D in such a way that the extensions belong to the kernel of any differential operator that commutes with the analytic automorphisms. We can define the Poisson operator by the formula [79]

$$(Pf)(z) = \int_{\partial D} (f \circ \omega_z)\, d\sigma, \tag{1}$$

where $\omega_z \in \pi^{-1}(z)$; since K acts transitively on ∂D, we can choose ω_z arbitrarily.

2.1. The adjective "Möbius", used in §2 of Chapter II, will be retained for $\operatorname{Aut} D$-invariant spaces, that is, spaces Y of functions in ∂D such that $f \circ \omega \in Y$ for any $f \in Y$ and every $\omega \in \operatorname{Aut} D$; this is immediate by the fact that the automorphisms of the classical domains (which will be considered below), in the matrix representation, are fractional-linear.

As usual, $A(\partial\Omega)$ denotes the algebra of boundary values (on the Shilov boundary) of functions holomorphic in D and continuous in the closure of D.

The following theorem states that $A(\partial D)$ is in a certain sense maximal in the class of all Möbius algebras.

THEOREM. *Let A be a closed Möbius subalgebra of $C(\partial D)$ such that $A(\partial D) \subset A$. Let δ_z denote the functional defined on $A(\partial D)$, the value of the holomorphic extension at the point $z \in D$ (e.g., the value of the Cauchy or Poisson integral at z). If there exists $z \in D$ such that δ_z can be extended to A as a multiplicative functional $\widetilde{\delta}_z$, then $A = A(\partial D)$.*

PROOF. Without loss of generality, we can assume that $z = 0$. Let μ be a probability measure on ∂D representing $\widetilde{\delta}_0$:

$$\widetilde{\delta}_0 f = \int_{\partial D} f\,d\mu, \qquad f \in A.$$

For $f \in A$ set

$$\widetilde{f}(z) = \int_K \widetilde{\delta}_0(f \circ \omega_z \circ k)\,dk = \int_{\partial D}\int_K (f \circ \omega_z \circ k)\,dk\,d\mu, \tag{2}$$

where ω_z is an arbitrary element of $\pi^{-1}(z)$. We can write (2)

$$\widetilde{f}(z) = \int_{\partial D} (f \circ \omega_z)\,d\nu,$$

where ν is some measure on ∂D. Clearly, ν is K-invariant and $\int_{\partial D} d\nu = 1$, therefore, $\nu = \sigma$. By formula (1), $\widetilde{f} = Pf$.

Since $\widetilde{\delta}_0$ is multiplicative, formula (2) implies that if $h \in A(\partial D)$, then $\widetilde{hf} = \widetilde{h}\widetilde{f}$. Hence

$$P(hf) = P(h)P(f).$$

Taking $h = z_k$, $k = 1, \dots, n$, we conclude that both Pf and $z_k Pf$ belong to the kernel of the invariant Laplace operator

$$\Delta = \sum_{i,j=1}^{n} g_{ij}(z)\frac{\partial^2}{\partial z_i \partial \overline{z}_j}.$$

Applying the Laplace operator, we get

$$\Delta(z_k Pf) = \sum_{i=1}^{n} g_{ki}(z)\frac{\partial}{\partial \overline{z}_k} Pf + z_k \Delta Pf = 0.$$

Since $\Delta Pf = 0$ and $g_{ij}(z)$ is a nonsingular matrix, this implies

$$\frac{\partial}{\partial \overline{z}_k} Pf = 0, \qquad k = 1, \dots, n.$$

Thus, Pf is the required holomorphic extension of f.

2.2. The following version of the above result will be useful.

THEOREM. *Let N be an analytic submanifold of D whose Shilov boundary ∂N is contained in ∂D. Let $f \in C(\partial D)$ be a function such that, for any analytic automorphism $\omega \in \mathrm{Aut}(D)$, the restriction $f \circ \omega|_{\partial N}$ has a continuous holomorphic extension to N. Then $f \in A(\partial D)$.*

The result follows immediately from Theorem 2.1, applied to the algebra of all functions satisfying the assumptions of the theorem. One defines $\widetilde{\delta}_z$ as the value at z of the holomorphic extension to N at the point $z \in D \cap N$.

REMARK. Suppose that D is the unit ball B^n in $\mathbb{C}^n$ and N is the intersection of B^n with some complex straight line. Since the automorphisms of the ball take straight lines into straight lines, it follows that the boundary values on the complex sphere of functions holomorphic in B^n are characterized by the existence of one-dimensional holomorphic extensions to sections of the ball by complex straight lines. Such a criterion for the existence of holomorphic extension, but in a stronger form—using the smaller number of complex straight lines—was proved by a different method in §2 of Chapter II. We shall return to this problem in Chapter IV.

2.3. In this section we characterize functions of $A(\partial D)$ in terms of the existence of analytic continuations into complex submanifolds of the topological boundary bD of a symmetric domain D.

The following notion is important in the geometry of symmetric domains [25]. Let D be a domain in $\mathbb{C}^n$ and bD its topological boundary. An analytic set $\varkappa \subset bD$ is called *a component* of (the boundary of) D if any holomorphic curve (holomorphic image of a disk in the complex plane) in bD, which is not disjoint from $\varkappa$, is contained in $\varkappa$.

A remarkable property of components of symmetric domains is that they are analytically equivalent to symmetric domains of fewer dimensions.

The components of the classical symmetric domains are listed in [25]. Linear components of the Siegel domains, as defined in §1, are components of the boundary in the sense of the above definition.

Note that, if all the components of the boundary of a symmetric domain have dimension zero, then the domain is analytically equivalent to a complex ball.

2.4. THEOREM. *Let D be a bounded symmetric domain in $\mathbb{C}^n$ not equivalent to a complex ball. Let $f \in C(\partial D)$ be a function such that, for any component $\varkappa$ of the boundary of D, the restriction $f|_{\partial\varkappa}$ to the Shilov boundary admits an analytic continuation to the component $\varkappa$. Then $f \in A(\partial D)$.*

PROOF. Suppose first that D is analytically equivalent to a Siegel domain $\Omega = \Omega_V$ of the first kind. Let $\gamma : \Omega \to D$ be a holomorphic diffeomorphism. We may assume that γ has a holomorphic extension to $\Omega \cup bD$. The set $\partial D \backslash \gamma(\partial\Omega)$, which corresponds to the points at infinity of the unbounded domain Ω, has measure zero on ∂D.

Consider an approximate unit

$$\alpha_k \in A(\partial\Omega) \cap C_0(\partial\Omega),$$

where α_k are uniformly bounded and converge pointwise to zero as $k \to \infty$ (cf. §1).

Transfer the function f to $\partial\Omega$ by setting $\widetilde{f} = f \circ \gamma$, and consider the functions $\widetilde{f}_k = \alpha_k f$. Then $\widetilde{f}_k \in C_0(\partial\Omega)$, and each of the functions $\widetilde{f}_k$, by assumption and by the analyticity of α_k, may be extended holomorphically to the components of the boundary of Ω. Since linear components are also components of the boundary, it follows that, for any face γ of the cone V, the functions $\widetilde{f}_k$ have analytic continuations into the manifolds $a + \Omega_\gamma$, $a \in \mathbb{R}^n$.

By Lemma 1.1, the Beurling spectrum $\sigma(\widetilde{f}_k)$ is contained in V^*. Since $V^* = \bigcap \gamma^*$ (the intersection extends over all faces γ), it follows that $\sigma(\widetilde{f}_k) \subset V^*$, and Lemma 1.2 implies $\widetilde{f}_k \in A(\partial\Omega)$. We now come back to the bounded realization. The functions $\widetilde{f}_k \circ \gamma^{-1}$ vanish on the "infinite" part $\partial D \backslash \gamma(\partial\Omega)$ of the Shilov boundary; we can therefore define $\widetilde{f}_k$ on ∂D by setting $f_k = \widetilde{f}_k \circ \gamma^{-1}$ on $\gamma(\partial\Omega)$ and zero on the rest of ∂D. Clearly, $f_k \in H^\infty(\partial D)$. It follows that $f \in A(\partial D)$. It is easy to justify these arguments. The sequence f_k is uniformly bounded and converges σ-a.e. to f. Since the algebra $A(\partial D)$ is K-invariant, its annihilator $[A(\partial D)]^\perp$ contains a dense subset of σ-absolutely continuous measures (cf. Lemma 3.1, Chapter I). The integrals of the functions f_k with respect to these measures vanish, so the integral of f also vanishes, since we can let $k \longrightarrow \infty$ under the integral sign. Then any measure from $[A(\partial D)]^\perp$ annihilates f, hence $f \in A(\partial D)$.

Suppose now that D has an unbounded realization as a Siegel domain of the second kind $\Omega = \Omega_{V,F}$. Consider the section $\Omega \cap \Pi$ of Ω by the complex plane $\Pi = \left\{(u, 0) \in \mathbb{C}^k \times \mathbb{C}^{n-k}\right\}$. Since $F(0,0) = 0$, $\Omega \cap \Pi$ is analytically equivalent to a Siegel domain of the first kind Ω_V over the same cone V. In addition, the Shilov boundary of the section is contained in that of the domain:

$$\partial(\Omega \cap \Pi) \subset \partial\Omega,$$

and the topological boundary $b_\Pi(\Omega \cap \Pi)$ relative to Π is contained in $b\Omega$. Referring to the definition of components, we readily see that each component of $\Omega \cap \Pi \cong \Omega_V$ is contained in a component of Ω. Since V is not one-dimensional, Ω_V has components of positive dimension.

In the bounded realization, the section $\Omega \cap \Pi$ becomes an analytic submanifold $N \subset D$ which is analytically equivalent to a bounded symmetric domain in $\mathbb{C}^k$ that has an unbounded realization as a Siegel domain of the first kind Ω_V. Since the components are compatible, the restriction $f|_{\partial N}$ has a holomorphic extension to the components of bN and, as proved, can be extended holomorphically to N. Analytic automorphisms map components onto components; hence, the same is true for translations $f\circ\omega$, $\omega\in \mathrm{Aut}(D)$. Theorem 2.2 now implies $f\in A(\partial D)$.

§3. Möbius spaces and algebras on Shilov boundaries of bounded symmetric domains

This section contains a partial description of the closed Möbius subspaces of $C(\partial D)$ and a complete description of the closed Möbius subalgebras of $C(\partial D)$ on the Shilov boundary of a symmetric domain D.

3.1. THEOREM. *Let D be a bounded irreducible symmetric domain in $\mathbb{C}^n$ which is a disk domain. Let Y be a closed Möbius subspace of $C(\partial D)$ (of $L^p(\partial D)$, $1\le p\le\infty$). Then one of the following assertions holds:*

(1) *Y is a subspace of $A(\partial D)$ (resp. of $H^p(\partial D)$).*

(2) *Y is a subspace of $\overline{A}(\partial D)$ (resp. of $\overline{H}^p(\partial D)$).*

(3) *Y contains nonconstant functions belonging to $A(\partial D)$ (resp. to $H^p(\partial D)$) and $\overline{A}(\partial D)$ (resp. $\overline{H}^p(\partial D)$).*

If Y contains nonconstant functions, then in case (1) *Y contains all the linear functions of $z_1, \dots, z_n$; in case* (2) *Y contains all the linear functions of $\overline{z}_1, \dots, \overline{z}_n$; in case* (3) *$Y$ contains all the linear functions of $z_1, \dots, z_n$; $\overline{z}_1, \dots, \overline{z}_n$.*

3.2. THEOREM. *Let D be a bounded irreducible symmetric domain in $\mathbb{C}^n$. A closed Möbius subalgebra of $C(\partial D)$ is one of the entries in the following diagram*

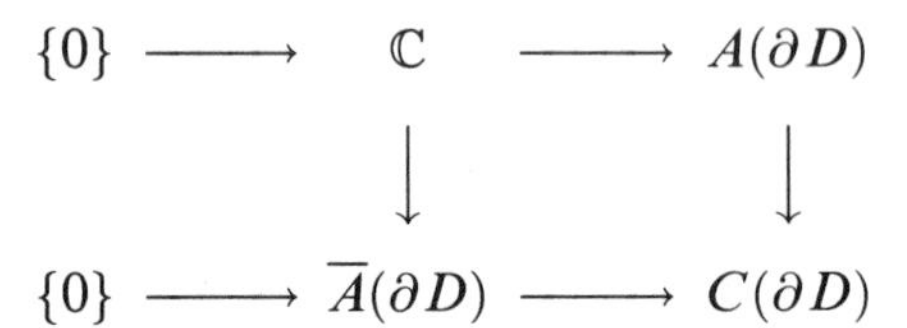

$$\begin{array}{ccccc} \{0\} & \longrightarrow & \mathbb{C} & \longrightarrow & A(\partial D) \\ & & \downarrow & & \downarrow \\ \{0\} & \longrightarrow & \overline{A}(\partial D) & \longrightarrow & C(\partial D) \end{array}$$

Theorem 3.2 is a straightforward corollary of Theorem 3.1. The second claim of Theorem 3.1, concerning linear functions, follows from the main assertion of that theorem and from the following lemma.

3.3. LEMMA. *Let $D\subset\mathbb{C}^n$ be a bounded symmetric domain which is a disk domain, and Y a closed Möbius subspace of $A(\partial D)$. If Y contains nonconstant functions, then it contains all the linear functions of $z_1, \dots, z_n$.*

PROOF. Extending the elements of Y holomorphically to D, one can consider Y as a subspace of $A(D)$. By Lemma 4.1, Chapter I, Y separates points of D, so there exist functions $f_1, \dots, f_N\in Y$ and a point $z_0\in D$ such that the rank of the Jacobian

$$\left(\frac{\partial f_i}{\partial z_j}\right)_{i,j=1}^{N,n}(z_0)$$

is maximal (see Lemma 2.3, Chapter I). Without loss of generality, we can assume that $z_0 = 0$ and that

$$\det\left(\frac{\partial f_i}{\partial z_j}(0)\right)_{i,j=1}^{n} \neq 0.$$

Consider the functions

$$\alpha_i(z) = \frac{1}{2\pi}\int_0^{2\pi} f_i(e^{i\theta} z)e^{-i\theta}\, d\theta,$$

which are obviously elements of Y. Since the functions α_i are holomorphic, we have

$$\alpha_i(z) = \sum_{j=1}^{n} \frac{\partial f_i}{\partial z_j}(0) z_j.$$

As the matrix of this linear system is nonsingular, the z_j are linear functions of α_i; hence, $z_j \in Y$.

3.4. Proof of Theorem 3.1. We shall use induction on the maximal dimension of the components of the boundary of the domain D. Note that, by Lemma 3.1, Chapter I, it suffices to consider the case $Y \subset C(\partial D)$.

The idea of the induction step is to consider projections of the space Y, each element being mapped onto its restrictions to the Shilov boundary of the components, using a retraction of the domain with the help of a one-parameter family of automorphisms.

In order to construct this family, let us represent the domain D as the Siegel domain of the third kind whose base is the component $\varkappa$ [25]:

$$S = \{(z, u, t) \in \mathbb{C}^k \times \mathbb{C}^l \times \varkappa : \operatorname{Im} z - \operatorname{Re} L_t(u, u) \in V,\ t \in \varkappa\},$$

where V is an open pointed convex cone in $\mathbb{R}^k$ and $t \to L_t$ a map defined on $\varkappa$ with values in the space of nondegenerate semihermitian forms L_t : $\mathbb{C}^l \times \mathbb{C}^l \to \mathbb{C}^k$. To the set of automorphisms of D that leave each point of $\varkappa$ fixed there correspond automorphisms of S including the transformations

$$z \to \alpha(z), \qquad u \to \beta(t)u, \quad t \to t,$$

where α is an affine transformation of the cone V, β an analytic matrix-valued function in $\varkappa$, and

$$L_t(\beta(t)u, \beta(t)v) = \alpha L_t(u, v)$$

(α is considered as an affine transformation of the space $\mathbb{C}^k$).

The set of these transformations contains a one-parameter family

$$\varphi_\lambda(z, u, t) = (\lambda^2 z, \lambda u, t).$$

Let ω_λ denote the corresponding automorphisms of D. As λ tends to zero, the maps φ_λ tend to the projection

$$(z, u, t) \to (0, 0, t).$$

Hence, on the "finite" part $\partial D \setminus \partial \varkappa_\infty$ of the Shilov boundary, where $\varkappa_\infty$ is the component corresponding to the component of S at infinity, the maps ω_λ tend to a map

$$\pi : \partial D \setminus \partial \varkappa_\infty \to \partial \varkappa$$

which is the identity on $\partial \varkappa$.

We can now prove the main statement of the theorem. Let $\chi(D)$ denote the maximal dimension of the components of the boundary of D. We will use induction on $\chi(D)$.

Induction base. If $\chi(D) = 0$, then D is biholomorphically equivalent to a complex ball, and an even stronger statement is true (see Chapter II, §2.6.4).

Induction step. Suppose the theorem is true for all domains D' with $\chi(D') < \chi(D)$. Let $\varkappa$ denote a component of the boundary of maximal dimension. Since analytic automorphisms of $\varkappa$ can be extended to automorphisms of the whole of D, it follows that the uniform closure

$$Y(\partial \varkappa) = \mathrm{cl}_C \left[Y|_{\partial \varkappa} \right]$$

of the restriction of Y to $\partial \varkappa$ is a Möbius space (we are identifying $\varkappa$ with an analytically equivalent bounded symmetric domain).

Since $\chi(\varkappa) < \chi(D)$, we can use the induction hypothesis. Therefore, one of the following conditions is fulfilled:

(1) $Y(\partial \varkappa) \subset A(\partial \varkappa)$,
(2) $Y \subset \overline{A}(\partial \varkappa)$,
(3) Y contains nonconstant functions from both $A(\partial \varkappa)$ and $\overline{A}(\partial \varkappa)$.

Suppose that either (1) or (2) holds. This means that all functions in Y have an analytic (resp. antianalytic) extension to the analytic manifold $\varkappa$ from its Shilov boundary. Since Y is a Möbius space, the functions in Y can be analytically extended into all the components of maximal dimension. All other components are components of the boundaries of components of the maximal dimension . Thus we can apply Theorem 2.4 to conclude that $Y \subset A(\partial D)$, resp. $Y \subset \overline{A}(\partial D)$.

Now suppose that (3) holds, and let $f \in Y(\partial \varkappa) \cap A(\partial \varkappa)$ be a nonconstant function. There exists a sequence $f_n \in Y$ converging to f uniformly on $\partial \varkappa$. Consider the family of automorphisms ω_λ constructed above. The function $f_n \circ \omega_\lambda$ belongs to Y, and for all $z \in \partial D \setminus \partial \varkappa_\infty$ we have

$$\lim_{\lambda \to 0} f(\omega_\lambda(z)) = f(\pi_\varkappa(z)).$$

Extend $f_n \circ \pi_\varkappa$ to $\partial \varkappa_\infty$ by setting $f_n \circ \pi_\varkappa|_{\partial \varkappa_\infty} = 0$. The function $f \circ \pi_\varkappa$ is equal almost everywhere in ∂D to the limit with respect to λ and then with respect to n of the uniformly bounded sequence of functions $f_n \circ \omega_\lambda$ in Y. Hence, if $\mu \in Y^\perp$ is an absolutely continuous measure, then

$$\int_{\partial D} (f \circ \pi_\varkappa)\, d\mu = 0.$$

The function $\widetilde{f} = f \circ \pi_{\varkappa}$ may be replaced by a continuous function on ∂D by taking its convolution with a function φ continous on K:

$$\varphi * \widetilde{f} = \int_K \varphi(k) T_k \widetilde{f}\, dk.$$

The set of absolutely continuous measures $\mu \in Y^{\perp}$ is weakly dense in Y, and therefore the function $\varphi * \widetilde{f}$ belongs to Y. It remains to note that, since $f \in A(\partial \varkappa)$, it follows that $\varphi * \widetilde{f} \in A(\partial D)$, and we can choose φ in such a way that $\varphi * \widetilde{f} \neq \text{const}$.

A similar argument proves that $Y \cap \overline{A}(\partial D) \not\subset \mathbb{C}$.

3.5. REMARK. Confining attention to the classical symmetric domains, i.e., excluding two exceptional types of symmetric domains in $\mathbb{C}^{16}$ and in $\mathbb{C}^{27}$, one can prove Theorem 3.1 more simply, without using unbounded representations of the domains. Let us sketch the construction of the retraction to the Shilov boundary of the component. This construction was basic in proving that the existence of a nonconstant function in $Y(\partial \varkappa) \cap A(\partial \varkappa)$ implies the existence of a nonconstant function in $Y \cap A(\partial D)$. Moreover, the explicit construction enables one to pick out polynomials of degree higher than one in Y.

Let D be a classical symmetric domain, i.e., a domain equivalent to one of the following domains:

$$D^{\mathrm{I}}_{n,m} = \left\{z \in \mathrm{Mat}(n \times m, \mathbb{C}) : 1 - zz^* > 0\right\},$$
$$D^{\mathrm{II}}_{n} = \left\{z \in \mathrm{Mat}(n, \mathbb{C}) : 1 - zz^* > 0,\ z^T = z\right\},$$
$$D^{\mathrm{III}}_{n} = \left\{z \in \mathrm{Mat}(n, \mathbb{C}) : 1 - zz^* > 0,\ z^T = -z\right\},$$
$$D^{\mathrm{IV}}_{n} = \{z \in \mathbb{C}^n : |z_1|^2 + |z_2|^2 + 2|z_3|^2 + \cdots + 2|z_n|^2 < 1 + |z_1 z_2 + z_3^2 + \dots z_n^2| < 2\},$$

where "> 0" means that the matrix is positive definite, z^* is the hermitian conjugate, and z^T is the transposed matrix.

We may suppose that $\varkappa$ is the standard component [25]

$$\varkappa = \begin{pmatrix} \mathbf{e}_r & 0 \\ 0 & w \end{pmatrix}$$

if $D = D^{\mathrm{I}}_{m,n}$ or $D = D^{\mathrm{II}}_{n}$, or

$$\varkappa = \begin{pmatrix} \mathbf{j}_r & 0 \\ 0 & w \end{pmatrix}$$

if $D = D^{\mathrm{III}}_{n}$. Here $\mathbf{e}_r$ is the $r \times r$ identity matrix,

$$\mathbf{j}_r = \begin{pmatrix} 0 & \mathbf{e}_r \\ -\mathbf{e}_r & 0 \end{pmatrix}.$$

Finally, if $D = D_n^{\mathrm{IV}}$, then

$$\varkappa = \left(1, z_2, 0, \dots, 0\right).$$

Define the automorphisms ω_λ as follows (see [30]):

$$\omega_\lambda(z) = (A_\lambda z + B_\lambda)(C_\lambda z + D_\lambda)^{-1}, \qquad \lambda \in [0, 1),$$

where the coefficients are given by

$$A_\lambda = \begin{pmatrix} (1-\lambda^2)^{-1/2}\mathbf{e}_r & 0 \\ 0 & \mathbf{e}_{m-r} \end{pmatrix}, \qquad B_\lambda = \begin{pmatrix} \lambda(1-\lambda^2)^{-1/2}\mathbf{e}_r & 0 \\ 0 & 0 \end{pmatrix},$$

$$C_\lambda = B_\lambda^{\mathrm{T}}, \qquad D_\lambda = \begin{pmatrix} (1-\lambda^2)^{-1/2}\mathbf{e}_r & 0 \\ 0 & \mathbf{e}_{n-r} \end{pmatrix}$$

if $D = D_{m,n}^{\mathrm{I}}$; if $D = D_n^{\mathrm{II}}$, set $n = m$. If $D = D_n^{\mathrm{III}}$, then

$$A_\lambda = D_\lambda = \begin{pmatrix} (1-\lambda^2)^{-1/2}\mathbf{e}_{2r} & 0 \\ 0 & \mathbf{e}_{n-2r} \end{pmatrix},$$

$$B_\lambda = C_\lambda = \begin{pmatrix} \lambda(1-\lambda^2)^{-1/2}\mathbf{e}_{2r} & 0 \\ 0 & 0 \end{pmatrix}.$$

Finally, if $D = D_n^{\mathrm{IV}}$, then

$$\omega_\lambda(z) = \frac{1}{i + \lambda z_1}\Big(z_1 + i\lambda,\ z_2 - i\lambda(z_1 z_2 + z_3^2 + \dots + z_n^2),\ (1-\lambda^2)^{-1/2} z_3, \\ \dots, (1-\lambda^2)^{-1/2} z_n\Big).$$

In the case of components $\varkappa_\infty$ "at infinity", $\omega_\lambda(z)$ does not tend to a limit as $\lambda \longrightarrow 1$, and the Shilov boundary of such components for domains of types I and II is defined by the condition $\det(\mathbf{e}_r + \mathbf{z}^{11}) = 0$, where

$$z = \begin{pmatrix} z^{11} & z^{12} \\ z^{21} & z^{22} \end{pmatrix},$$

z^{11} is a $r \times r$ matrix. For domains of type III these components are defined by the condition $\det(\mathbf{e}_{2r} + z^{11}) = 0$, where z^{11} is the $2r \times 2r$ matrix. For domains of type IV the condition is $z_1 + i = 0$.

The limit map $\pi_\varkappa = \lim_{\lambda\to 1} \omega_\lambda$ for domains of types I, II is

$$\pi_\varkappa = \begin{pmatrix} \mathbf{j}_r & 0 \\ 0 & \mathbf{u}_r(z) \end{pmatrix}, \qquad \mathbf{u}_r(z) = z^{22} - z^{21}(\mathbf{e}_r + z^{11})^{-1} z^{12}.$$

For domains of type III,

$$\pi_\varkappa = \begin{pmatrix} \mathbf{e}_r & 0 \\ 0 & \mathbf{u}_r(z) \end{pmatrix}, \qquad \mathbf{u}_r(z) = z^{22} - z^{21}(\mathbf{e}_{2r} + z^{11})^{-1} z^{12}.$$

Finally, for domains of type IV,

$$\pi_\varkappa = \left(1, -iz_2 + (i + z_1)^{-1}(z_3^2 + \dots + z_n^2), 0, \dots, 0\right).$$

Suppose now that we define the function $f \in Y$, $f \in Y$, $f|_{\partial\varkappa} \in A(\partial\varkappa)$ in the proof of Theorem 3.1 as the projection onto the matrix elememt corresponding to the block w in the representation of $\varkappa$. Then $f \circ \pi_\lambda = \lim_{\lambda\to\infty} f \circ \omega_\lambda$ coincides on $\partial D \setminus \partial\varkappa_\infty$ with an element of the matrix $\mathbf{u}_r(z)$. Since the elements of the matrix

$$\mathcal{J}(z) = \frac{1}{2\pi}\int_0^{2\pi} \mathbf{u}_r(z)e^{-i\varphi}\,d\varphi,$$

which is z^{22} for domains of types I–III and $z_1(z_3^2 + \cdots + z_n^2)$ for domains of type IV, belong to Y by construction, this proves that $Y \cap A(\partial D) \not\subset \mathbb{C}$.

3.6. This section is devoted to a description of the closed Möbius algebras of continuous functions on reducible bounded symmetric domains.

Let D be a symmetric domain, and consider its decomposition as a direct product of irreducible symmetric domains:

$$D = D_1' \times \cdots \times D_M'. \tag{1}$$

Grouping together analytically equivalent factors, we can express D as a direct product $D = D_1 \times \cdots \times D_N$ of a maximal number of mutually inequivalent symmetric domains.

Let P, Q be two arbitrary sets of positive integers not greater than N. Let $A_{P,Q}(\partial D)$ denote the algebra of functions in $C(\partial D)$ which have continuous holomorphic extensions to D_i as functions of $z_i \in \partial D_i$, $i \in P$, and a continuous antiholomorphic extensions to D_j as functions of $z_j \in \partial D_j$, $j \in Q$, all other variables remaining fixed. The case $P = Q = \varnothing$ corresponds to the algebra $C(\partial D)$; if $P = \{1, \dots, N\}$, $Q = \varnothing$ and $P = \varnothing$, $Q = \{1, \dots, N\}$, one gets the algebras $A(\partial D)$ and $\overline{A}(\partial D)$, respectively. The functions belonging to $A_{P,Q}(\partial D)$ are constant in the variables z_i, $i \in P \cap Q$.

Clearly the algebras $A_{P,Q}(\partial D)$ are invariant under the analytic automorphisms of D. We are going to prove that there are no other closed Möbius subalgebras in $C(\partial D)$. Let us first prove the following assertion.

3.7. Lemma. *Let D^1, D^2 be bounded symmetric domains, $D = D^1 \times D^2$, and A a closed Möbius subalgebra of $C(\partial D)$. Let A^1 denote the subalgebra of all functions in A that are constant as functions of $z_2 \in D_2$ for each fixed $z_1 \in \partial D^1$. Consider A^1 as an algebra of functions on ∂D^1. Then, for every $z_2 \in \partial D^2$, the restriction of A to the fiber $\partial D^1 \times \{z_2\}$ coincides with A^1:*

$$A|_{\partial D^1 \times \{z_2\}} = A^1.$$

Proof. The argument uses the fact that averaging with respect to a fixed variable over the stationary subgroup of automorphisms at some interior point becomes the restriction when the point approaches the Shilov boundary.

Pick a point $w_2 \in D^2$. Let $f \in A$. Consider the averaging operator

$$(S_{w_2} f)(z) = \int_{\mathrm{Aut}(D^2, w_2)} f(z_1, , kz_2)\, dk, \qquad z = (z_1, z_2) \in \partial D,$$

where $\mathrm{Aut}(D^2, w_2)$ is the stationary subgroup at w_2. As this group acts transitively on ∂D^2, the function $S_{w_2} f$ depends only on z_1. Since $S_{w_2} f \in A$, we have $S_{w_2} f \in A^1$. The operator S_{w_2} is simply the invariant Poisson integral P_{D^2} corresponding to the domain D^2 (see §2, formula (1)):

$$(S_{w_2} f)(z_1, z_2) = (P_{D^2} f_{z_1})(w_2),$$

where $f_{z_1} = f(z_1, \cdot)$. As w_2 tends to a point w_2^0 on the Shilov boundary,

$$\lim_{w_2 \to w_2^0} (S_{w_2} f)(z_1, z_2) = \lim_{w_2 \to w_2^0} (P_{D^2} f_{z_1})(w_2) = f(z_1, w_2^0).$$

Thus the function $\tilde{f}(z_1, z_2) = f(z_1, w_2^0)$ is the pointwise limit on ∂D of a sequence of functions belonging to A^1. Since this sequence is uniformly bounded and the limit is a continuous function, a standard argument, using orthogonal absolutely continuous measures (see, for example, the proof of Theorem 2.4), will show that $\tilde{f} \in A^1$. Thus, the restriction of f to the fiber $\partial D \times \{w_2^0\}$, extended to ∂D as a constant function of $z_2 \in \partial D^2$, yields an element $\tilde{f} \in A^1$. Therefore, $A|_{\partial D^1 \times \{w_2^0\}} \subset A^1$. The reverse inclusion is trivial.

3.8. THEOREM. *Let D be a bounded symmetric domain. Then the list of all closed Möbius subalgebras of $C(\partial D)$ consists of $\{0\}$ and the algebras $A_{P,Q}(\partial D)$.*

PROOF. Let A be a closed Möbius subalgebra of $C(\partial D)$, $A \neq \{0\}$. Consider the decomposition (1) of §3.6, and for any $i = 1, \dots, M$ let A_i' denote the subalgebra of A consisting of the functions that depend only on $z_i \in \partial D_i'$ and are constant in the other variables. By Theorem 3.2, A_i' is one of the algebras $\{0\}$, $\mathbb{C}$, $A(\partial D_i')$, $\overline{A}(\partial D_i')$, $C(\partial D_i')$. The case $A_i' = \{0\}$ is impossible by Lemma 3.7, because by assumption $A \neq \{0\}$.

The domain D is analytically equivalent to a domain for which the analytically equivalent factors in (1) are identical. We may assume that D is such a domain. In that case permutations of the coordinates corresponding to identical domains are analytic automorphisms of D. Since A is invariant under such permutations, each of the algebras A_i' that correspond to equivalent factors in (1) must be the same item in the above list.

Now consider the decomposition in which the equivalent factors are grouped together. We see that for any $i = 1, \dots, N$ the algebra A_i, defined as the set of functions in A that are constant in all the variables except

$z_i \in \partial D_i$, is one of $\mathbb{C}$, $A(\partial D_i)$, $\overline{A}(\partial D_i)$, $C(\partial D_i)$ (note that the existence of holomorphic extention of a direct product from the Shilov boundary is equivalent to the existence of holomorphic extension with respect to the respective variables of the factors).

Replace z_j by $\overline{z}_j$ for all j such that $A_j = \overline{A}(\partial D_j)$. Then the situation is as follows. The domain D may be represented as a direct product

$$D = D^1 \times D^2 \times D^3$$

of three symmetric domains, and the algebras A^i of functions in A that depend only on $z_i \in \partial D^i$ are $A^1 = A(\partial D^1)$, $A^2 = C(\partial D^2)$, $A^3 = \mathbb{C}$. By Lemma 3.7, all the functions in A are constant in $z_3 \in \partial D^3$, so we can consider A as a function algebra on $\partial D^1 \times \partial D^2$. To complete the proof, we will use Bishop's theorem on maximal antisymmetry sets (MAS) (see §2.1, Chapter I). It follows from the condition $A^2 = C(\partial D^2)$ that every MAS is contained in some fiber $\partial D^1 \times \{z_2\}$, $z_2 \in \partial D^2$. Since $A^1 = A(\partial D^1)$, it follows from Lemma 3.7 that the restriction of A to these fibers coincides with the restriction to the same fibers of the algebra of all functions in $C(\partial D^1 \times \partial D^2)$ that admit continuous holomorphic extensions to D^1 as functions of $z_1 \in \partial D^1$ for each fixed $z_2 \in \partial D^2$. By Bishop's theorem, an algebra is uniquely determined by its restrictions to MAS; hence, the two algebras coincide. This completes the proof.

§4. Pseudounitarily invariant spaces of analytic functions on Stiefel manifolds

In the preceding section we obtained a complete description of the byholomorphic invariant algebras on Shilov boundaries of bounded symmetric domains. As for function spaces, some of their properties were established. In the present section we confine ourselves to spaces of boundary values of analytic functions on the Shilov boundaries of symmetric domains of type I. In this situation it is possible to list all such spaces. Our approach is apparently also applicable to the description of invariant spaces of continuous or integrable functions. In addition, it is probably possible to avoid using complex coordinates and to extend the argument to arbitrary symmetric domains, relying on Cartan's general construction.([1])

4.1. Introduction. The Shilov boundary of the classical symmetric domain $D_{n,m}$ of type I is the Stiefel manifold

$$St_{m,n} = U(n)/U(n-m),$$

considered as the manifold of all $m \times n$ complex matrices, $m \leq n$, that satisfy the condition

$$Z^* Z = 1_n.$$

([1]) *Remark for the English edition*: See [116], [117].

The group of analytic automorphisms of $D_{n,m}$ is isomorphic to the special pseudounitary group $SU(m, n)$, i.e., the group of matrices $\begin{pmatrix} A & B \\ C & D \end{pmatrix}$ with determinant one, where the blocks A, B, C, D are $m \times m$, $m \times n$, $n \times m$, and $n \times n$ matrices, respectively, that leave the quadratic form

$$-|z_1|^2 - \cdots - |z_m|^2 + |z_{m+1}|^2 + \cdots + |z_{m+n}|^2$$

invariant. The group $SU(m, n)$ acts on $St_{m,n}$ as the group of Möbius transformations:

$$\omega(Z) = (AZ + B)(CZ + D)^{-1}. \tag{1}$$

We can extend this action to $\overline{D}_{n,m}$ by continuity, and the group $SU(m, n)$ acts transitively on the Stiefel manifold. The stationary subgroup of the zero element, $K \subset \operatorname{Aut} D_{n,m}$, which is isomorphic to the maximal compact subgroup $S(U(m) \times U(n)) \subset SU(m, n)$, acts on $St_{m,n}$ (again transitively) by two-sided unitary translations:

$$S(U(m) \times U(n)) \ni (A, B) : Z \to A^{-1}ZB.$$

There exists a normalized K-invariant measure σ on $St_{m,n}$. The corresponding Lebesgue spaces will be denoted by $L^p(St_{m,n})$. Let $C(St_{m,n})$ denote the space of continuous complex-valued functions in $St_{m,n}$ endowed with the uniform norm.

We will consider the translation operators T_ω, $\omega \in SU(m, n)$, in spaces of functions defined on $St_{m,n}$: $T_\omega f = f \circ \omega^{-1}$. Call a space W of functions on $St_{m,n}$ *pseudounitarily invariant* (*$SU(m, n)$-invariant,* or simply *invariant*) if $T_\omega f \in W$ for any $f \in W$ and $\omega \in SU(m, n)$. Let $A(St_{m,n})$, $H^p(St_{m,n})$, $1 \leq p \leq \infty$, denote the closures of the spaces of complex polynomials in the matrix elements z_{ij} with respect to the norms of $C(St_{m,n})$, $L^p(St_{m,n})$, respectively. That these spaces are invariant is obvious.

Our problem is: Do these spaces have closed invariant subspaces?

When $m = 1$, the Stiefel manifold $St_{m,n}$ coincides with the unit sphere S^{2n-1} in $\mathbb{C}^n$, and the answer is known [84] (see also Rudin's monograph [26]): There are no proper closed invariant subspaces of $A(St_{m,n})$ and $H^p(St_{m,n})$ other than $\{0\}$ and $\mathbb{C}$. The following example shows that when $m > 1$ there do exist nontrivial proper subspaces.

EXAMPLE. Let $m = n > 1$. Then the Stiefel manifold $St_{m,n}$ is just the group $U(n)$ of unitary matrices. Let V_s denote the minimal $SU(n, n)$-invariant closed subspace of $H^2(St_{n,m})$ that contains the function $\det^s Z$. We claim that, for any $s = 0, 1, \ldots, n-1$, V_s is a proper subspace of $H^2(St_{n,m})$.

To that end, it suffices to show that $\det^n Z$ is not in any of the spaces V_s, $0 \leq s \leq n-1$.

Pick $\omega \in SU(n, n)$, and consider the scalar product

$$I = \langle T_\omega \det^s, \det^n \rangle = \int_{U(n)} \det^s[(AZ+B)(CZ+D)^{-1}]\overline{\det^n(Z)}\, d\sigma(Z).$$

Change the variable of integration as follows:

$$W = (AZ+B)(CZ+D)^{-1}, \qquad Z = (A - WC)^{-1}(WD - B).$$

The real Jacobian of this transformation is the invariant Poisson kernel $P(\omega(0), W)$, where

$$P(X, W) = \frac{\det^n(1 - XX^*)}{|\det(X - W)|^{2n}}.$$

Taking into account that $\omega(0) = BD^{-1}$, $WW^* = 1$, we get

$$I = \text{const} \int_{U(n)} \det^{n-s}(W^*)\det^{-n}[(1 - (A^*)^{-1}C^*W^*)(1 - BD^{-1}W^*)]\, d\sigma(W).$$

The integrand is an antiholomorphic function in $D_{n,n}$. Therefore, the integral equals the value of the integrand at $W = 0$ and vanishes when $n-s > 0$. Thus $V_s \neq H^2(St_{n,n})$, $0 \le s \le n$.

It will be shown below that all the invariant spaces of holomorphic functions on the Stiefel manifolds can be constructed in this way.

We now formulate the main result. Let $\det_j(Z)$ denote the jth principal minor of Z:

$$\det_j(Z) = \det(z_{\alpha\beta})_{\alpha,\beta=1}^{j,j}.$$

Let X be one of the spaces $A(St_{m,n})$, $H^p(St_{m,n})$, $1 \le p < \infty$. Let $\mathrm{cl}_X[\det_m^S]_{SU(u,m)}$ denote the closed $SU(m, n)$-cyclic span (in X) of the function $\det_m^s$ by V_s.

4.1.1. THEOREM. *A closed $SU(n, m)$-invariant subspace of X is one of the following:*

$$\{0\} = V_0 \subset V_1 \subset V_2 \subset \cdots \subset V_{m-1} \subset V_m = X.$$

The space V_s is the closed $S(U(m) \times U(n))$-cyclic span of the highest weight vectors

$$\varphi_{\overline{k}}(Z) = \prod_{j=1}^{m} \det_j^{k_j - k_{j+1}}(Z),$$

where $\overline{k} = (k_1, \dots, k_m)$ runs through the set of signatures $k_1 \ge \cdots \ge k_m \ge k_{m+1} = 0$ such that $k_{s+1} \le s$.

4.2. Infinitesimal action of the group $U(m, n)$. Every $SU(m, n)$-invariant space is also $S(U(m)\times U(n))$-invariant and can therefore be decomposed as a sum of irreducible representations of the group $K \cong S(U(m) \times U(n))$. Thus the problem of describing the $SU(m, n)$-invariant spaces reduces to finding families of irreducible representations of K that are *stable* under the action of the ambient pseudounitary group. A family of representations is said to be stable if each space of an irreducible representation in the family is mapped by a translation operator in $SU(m, n)$ onto a space whose irreducible components all belong to the same family. In order to describe families with this property, we have to study the infinitesimal action of the pseudounitary group on the irreducible components of its maximal compact subgroup.

By formula (1), the groups $U(m, n)$ and $SU(m, n)$ generate the same set of analytic automorphisms. Instead of $SU(m, n)$, therefore, we may consider the more convenient group $U(m, n)$.

The standard basis of the Lie algebra $\mathfrak{u}(m, n)$ of $U(m, n)$:

$$a_{ij} = e_{ij} - e_{ji}, \qquad b_{ij} = \sqrt{-1}(e_{ij} + e_{ji}),$$

$i \neq j$ and either $1 \leq i, j \leq m$ or $m+1 \leq i, j \leq m+n$;

$$c_i = \sqrt{-1}e_{ii}, \qquad 1 \leq i \leq m+n;$$

$$d_{ij} = e_{ij} + e_{ji}, \quad f_{ij} = \sqrt{-1}(e_{ij} - e_{ji}), \qquad 1 \leq i \leq m, \ m+1 \leq j \leq m+n.$$

Here e_{ij} is the $(m+n)\times(m+n)$-matrix all of whose elements are zero except the (i, j) element, which is one.

The elements a_{ij}, b_{ij}, c_i form a basis of the Lie algebra $\mathfrak{k}$ of K. The Cartan decomposition is

$$\mathfrak{u}(m, n) = \mathfrak{k} + \rho,$$

where ρ is the complementary space with basis d_{ij}, f_{ij}. There is a similar decomposition for the complex algebras:

$$\mathfrak{u}^{\mathbb{C}}(m, n) = \mathfrak{k}^{\mathbb{C}} + \rho^{\mathbb{C}},$$

where $\mathfrak{k}^{\mathbb{C}}$ is the complex space with basis

$$l_{ij} = \frac{1}{2}(a_{ij} + \sqrt{-1}b_{ij}), \qquad 1 \leq i \neq j \leq m;$$

$$r_{ij} = \frac{1}{2}(-a_{i+m, j+m} + \sqrt{-1}b_{i+m, j+m}), \qquad 1 \leq i \neq j \leq n;$$

$$l_{ii} = \sqrt{-1}c_i, \quad r_{ii} = \sqrt{-1}c_{i+n}, \qquad 1 \leq i \leq m;$$

and $\rho^{\mathbb{C}}$ is the space with basis

$$t_{kl} = \frac{1}{2}(d_{k, l+m} + \sqrt{-1}f_{k, l+m}),$$

$$\bar{t}_{kl} = \frac{1}{2}(d_{k, l+m} - \sqrt{-1}f_{k, l+m}), \qquad 1 \leq k \leq m, \ 1 \leq l \leq n.$$

Let ρ^+ denote the linear span of the vectors t_{kl} and ρ^- the linear span of the vectors $\bar{t}_{kl}$. Then

$$\mathfrak{u}^{\mathbb{C}}(m, n) = \mathfrak{k}^{\mathbb{C}} + \rho^+ + \rho^- .$$

The following commutation relations hold:

$$\left[\rho^{\pm}, \rho^{\pm}\right] = 0, \qquad \left[\rho^{\pm}, \rho^{\mp}\right] = \mathfrak{k}^{\mathbb{C}}, \qquad \left[\mathfrak{k}^{\mathbb{C}}, \rho^{\pm}\right] = \rho^{\pm}.$$

Consider the representation of the algebra $\mathfrak{u}^{\mathbb{C}}(m, n)$ in the Lie algebra of differential operators on $C^{\infty}(St_{m,n})$. Let $\mathfrak{g}$ denote the Lie algebra generated by the differential operators $D_{a_{ij}}, \dots, D_{f_{ij}}$ corresponding to the basis elements of $\mathfrak{u}(m, n)$. Define a natural basis of the corresponding complex algebra $\mathfrak{g}^{\mathbb{C}}$:

$$\begin{aligned}
L_{ij} &= \frac{1}{2}(D_{a_{ij}} + \sqrt{-1}D_{b_{ij}}) = Z_i\nabla_j - \overline{Z}_j\overline{\nabla}_i, \\
L_{ii} &= -\sqrt{-1}D_{c_i} = Z_i\nabla_i - \overline{Z}_i\overline{\nabla}_i; \\
R_{ij} &= \frac{1}{2}(-D_{a_{i+m,j+m}} + \sqrt{-1}D_{b_{i+m,j+m}}) = Z^i\nabla_j - \overline{Z}^j\overline{\nabla}_i, \\
R_{ii} &= -\sqrt{-1}D_{c_{i+m}} = Z^i\nabla_i - \overline{Z}^i\overline{\nabla}_i; \\
T_{ij} &= \frac{1}{2}(D_{d_{i,j+m}} + \sqrt{-1}D_{f_{i,j+m}}) = \frac{\partial}{\partial \bar{z}_{ij}} - \sum_{k,l=1}^{m,n} z_{kl}z_{il}\frac{\partial}{\partial z_{kl}}, \\
\overline{T}_{ij} &= \frac{1}{2}(D_{d_{i,j+m}} - \sqrt{-1}D_{f_{i,j+m}}) = \frac{\partial}{\partial z_{ij}} - \sum_{k,l=1}^{m,n} \bar{z}_{kl}\bar{z}_{il}\frac{\partial}{\partial \bar{z}_{kl}}.
\end{aligned} \tag{2}$$

Here Z_i, Z^j are the ith row and the jth column of Z, respectively;

$$\nabla^j = \left(\frac{\partial}{\partial z_{1j}}, \dots, \frac{\partial}{\partial z_{mj}}\right), \qquad \nabla_i = \left(\frac{\partial}{\partial z_{i1}}, \dots, \frac{\partial}{\partial z_{in}}\right).$$

The Cartan decomposition corresponds to the following decomposition of the Lie algebra of differential operators:

$$\mathfrak{g} = \mathfrak{k}^{\mathbb{C}} + \rho^+ + \rho^- ,$$

where $\mathfrak{k}^{\mathbb{C}}$ is the Lie algebra generated by the operators L_{ij}, R_{ij}; the spaces ρ^+ and ρ^- are generated by the operators T_{ij} and $\overline{T}_{ij}$, respectively.

4.2.1. PROPOSITION. *The adjoint operators satisfy the following relations*:

$$T^*_{kl} = -\overline{T}_{kl} + n\bar{z}_{kl}, \qquad \overline{T}^*_{kl} = -T_{kl} + nz_{kl}, \qquad L^*_{ij} = L_{ji}, \qquad R^*_{ij} = R_{ji}.$$

PROOF. The differential operators T_{kl} and $\overline{T}_{kl}$ correspond to the one-parameter subgroups

$$d_{k,l+m}(t) + \sqrt{-1}f_{k,l+m}(t), \qquad d_{k,l+m}(t) - \sqrt{-1}f_{k,l+m}(t),$$

where

$$d_{k,l+m}(t) = \frac{1}{\sqrt{1-t^2}}(e_{k,k} + e_{l+m,l+m} + td_{k,l+m}) + \sum_{\alpha\neq k,l+m} e_{\alpha\alpha},$$

$$f_{k,l+m}(t) = \frac{1}{\sqrt{1-t^2}}(e_{k,k} + e_{l+m,l+m} + tf_{k,l+m}) + \sum_{\alpha\neq k,l+m} e_{\alpha\alpha}.$$

Let $f, g \in C^\infty(St_{m,n})$. Then

$$\langle T_{kl}f, g\rangle = \frac{1}{2}\frac{d}{dt}\bigg|_{t=0}\left[\int_{St_{m,n}} f\circ d_{k,l+m}(t)\overline{g}\,d\sigma(Z)\right.$$

$$\left. + \sqrt{-1}\int_{St_{m,n}} f\circ f_{k,l+m}(t)\overline{g}\,d\sigma(Z)\right]. \quad (3)$$

Changing variables $Z = d_{i,j+m}(-t)(W)$ in the first integral and $Z = f_{i,j+m}(-t)(W)$ in the second, we obtain

$$\langle T_{kl}f, g\rangle = \frac{1}{2}\left[-\int_{St_{m,n}} f\frac{d}{dt}\bigg|_{t=0}\overline{g\circ d_{k,l+m}(t)}\,d\sigma(W)\right.$$

$$\left. + \sqrt{-1}\int_{St_{m,n}} f\frac{d}{dt}\bigg|_{t=0}\overline{g\circ f_{k,l+m}(t)}\,d\sigma(W)\right]$$

$$+\frac{1}{2}\int_{St_{m,n}} f\overline{g}\frac{d}{dt}\bigg|_{t=0}\left[\frac{(1-t^2)^n}{|1-\overline{w}_{k,l}t|^{2n}} + \frac{(1-t^2)^n}{|1-\sqrt{-1}\overline{w}_{k,l}t|^{2n}}\right]d\sigma(W)$$

$$= -\langle f, \overline{T}_{kl}g\rangle + n\int_{St_{m,n}} f\overline{g}w_{ij}\,d\sigma(W).$$

We have used the fact that after the change of variables in (3) there appears a third factor—the invariant Poisson kernel.

The other equalities are proved in a similar way.

4.3. Representations of the Lie algebra $\mathfrak{k}$. The following construction will be useful later.

Let π denote the quasi-regular representation of K in $H^2(St_{m,n})$:

$$\pi(k)f = T_kf = f\circ k^{-1},$$

and let $d\pi : X \to D_X$ denote the differential of π. We can extend $d\pi$ by linearity to a representation $d\pi^{\mathbb{C}}$ of the complexification $\mathfrak{k}^{\mathbb{C}}$ of the Lie algebra of K.

The factorization of π into irreducible representations is well known (see, for example, [77]). The space $H^2(St_{m,n})$ is a direct sum

$$H^2(St_{m,n}) = \bigoplus H_{\overline{k}}$$

of irreducible subspaces $H_{\overline{k}}$. The latter are indexed by m-tuples $\overline{k} = (k_1, \dots, k_m)$ of nonnegative integers in descending order (signatures). Each space $H_{\overline{k}}$ is the closed K-cyclic span of the highest weight vectors

$$\varphi_{\overline{k}}(Z) = \prod_{j=1}^{m} \det_j^{k_j - k_{j+1}}(Z), \qquad k_{m+1} = 0.$$

Each $H_{\overline{k}}$ is the space of an irreducible representation of K.

Let $\mathrm{ad}^P_{\mathfrak{k}^{\mathbb{C}}}$ denote the restriction of the adjoint representation of $\mathfrak{k}^{\mathbb{C}}$ to the ring $P[T]$ of polynomials in the commuting operators T_{ij}:

$$\mathrm{ad}^P_{\mathfrak{k}^{\mathbb{C}}}(X)(p) = [X, p], \qquad X \in \mathfrak{k}^{\mathbb{C}}, \ p \in P[T].$$

Let $d\pi^{\mathbb{C}}_P$ denote the restriction

$$d\pi^{\mathbb{C}}_P(X) = d\pi^{\mathbb{C}}(X)|_{P[Z]}, \qquad X \in \mathfrak{k}^{\mathbb{C}},$$

of the representation $d\pi^{\mathbb{C}}$ to the ring $P[Z]$ of polynomials in elements of Z.

It is easy to see that the representations $d\pi^{\mathbb{C}}_P$ and $\mathrm{ad}^P_{\mathfrak{k}^{\mathbb{C}}}$ are equivalent. Indeed, the equalities

$$L_{ks}(z_{ij}) = \delta_{si} z_{kj}, \qquad R_{ks}(z_{ij}) = \delta_{si} z_{ki}$$

simply duplicate the commutation relations among the operators L_{ks}, R_{ks}, T_{ij}:

$$[L_{ks}, T_{ij}]\delta_{si} T_{ki}, \qquad [R_{ks}, T_{ij}] = \delta_{si} T_{ki}.$$

So

$$[X, p(T)] = D_X(p)(T), \qquad p(T) \in P[T], \ X \in \mathfrak{k}^{\mathbb{C}}$$

(this is readily verified for the basis elements L_{ks}, R_{ks}; then use linearity). Thus, the operator

$$S : \ p(z_{ij}) \to p(T_{ij})$$

is an intertwining operator:

$$\left(\mathrm{ad}^P_{\mathfrak{k}^{\mathbb{C}}}(X) S\right)(p) = [X, p(T)] = D_X(p)(T) = S\, d\pi^{\mathbb{C}}_P(X)(p),$$

$$p \in P[Z], \ X \in \mathfrak{k}^{\mathbb{C}}.$$

Any polynomial $p(Z)$ may be represented as a finite linear combination of translations $T_u \varphi_{\overline{r}}$, $u \in K$, of highest weight vectors. Furthermore, if $u = \exp(X)$, $X \in \mathfrak{k}$, then it follows from the Taylor formula that

$$T_u \varphi_{\overline{r}} = \exp(D_X)\varphi_{\overline{r}} = \sum_{j=0}^{N} \frac{1}{j!} D_X^j \varphi_{\overline{r}}.$$

Hence

$$(T_u\varphi_{\bar r})(T) = \sum_{j=0}^{N} \frac{1}{j!}[X, \varphi_{\bar r}(T)]^{(j)},$$

where $[X, \varphi_{\bar r}(T)]^{(j)}$ denotes the multiple commutator

$$[X, \varphi_{\bar r}(T)]^{(j)} = \underbrace{[X, [X, \dots, [X,}_{j \text{ times}} \varphi_{\bar r}(T)] \dots].$$

Thus, any polynomial $p(T)$ may be represented as a linear combination of multiple commutators:

$$p(T) = \sum_{\bar r, j} \lambda_{\bar r, j}[X_{\bar r}, \varphi_{\bar r}(T)]^{(j)}, \qquad X_{\bar r} \in \mathfrak{k}^{\mathbb{C}}, \ \lambda_{\bar r, j} \in \mathbb{C}. \tag{4}$$

4.4. Operators increasing and decreasing signatures. We introduce the following notation for minors:

$$\Delta_{i_1, \dots, i_l}^{j_1, \dots, j_l}(Z) = \det\left(z_{i_\alpha, j_\beta}\right)_{\alpha, \beta=1}^{l, l};$$

$$\Delta_{i_1, \dots, i_l} = \Delta_{j_1, \dots, j_l}^{1, \dots, l}, \qquad \Delta^{j_1, \dots, j_l} = \Delta_{1, \dots, l}^{j_1, \dots, j_l}.$$

Using this notation,

$$\det_j = \Delta^{1, \dots, j} = \Delta_{1, \dots, j}.$$

The row and column expansions yield

$$\begin{aligned} &\sum_{\alpha=1}^{m} (-1)^{\alpha+\beta} z_{i_\alpha, k} \Delta_{i_1, \dots, \hat{i}_\alpha, \dots, i_l}^{j_1, \dots, \hat{j}_\beta, \dots, j_l} = \Delta_{i_1, \dots, i_l}^{j_1, \dots, j_{\beta-1}, k, j_{\beta+1}, \dots, j_l}, \\ &\sum_{\beta=1}^{n} (-1)^{\alpha+\beta} z_{k j_\beta} \Delta_{i_1, \dots, \hat{i}_\alpha, \dots, i_l}^{j_1, \dots, \hat{j}_\beta, \dots, j_l} = \Delta_{i_1, \dots, i_{\alpha-1}, k, i_{\alpha+1}, \dots, i_l}^{j_1, \dots, j_l}, \end{aligned} \tag{5}$$

where $\hat{i}_\alpha$ means that the index i_α is omitted. Set

$$I(i_1, \dots, i_p; q) = \sum_{\alpha=1}^{p} (-1)^{p+\alpha} \Delta_{i_1, \dots, \hat{i}_\alpha, \dots, i_l} T_{i_\alpha, p}(\det_q),$$

$$J(i_1, \dots, i_p) = \sum_{s=1}^{p} (-1)^{p+s} T_{i_s, p}\left(\Delta_{i_1, \dots, \hat{i}_s, \dots, i_l}\right).$$

4.4.1. LEMMA.

(i) $$T_{kl}\left(\Delta_{i_1,\dots,i_p}\right) = -\sum_{\beta=1}^{p} z_{k\beta}\Delta^{1,\dots,\beta-1,l,\beta+1,\dots,l}_{i_1,\dots,i_p}$$
$$= -\sum_{\alpha=1}^{p} z_{i_\alpha,l}\Delta_{i_1,\dots,i_{\alpha-1},k,i_{\alpha+1},\dots,i_p};$$

(ii) $T_{kl}(\det_p) = -z_{kl}\det_p$ *for* $\min(k, l) \le p$;

(iii) $$I(i_1, \dots, i_p; q) = \begin{cases} 0, & p > q, \\ -\Delta_{i_1,\dots,i_p}\cdot\det_q, & p \le q; \end{cases}$$

(iv) $J(i_1, \dots, i_p) = (p-1)\Delta_{i_1,\dots,i_p}$.

PROOF. (i) Using the definition (2) of T_{kl}, we write

$$T_{kl}\left(\Delta_{i_1,\dots,i_p}\right) = -\sum_{\gamma,\beta=1}^{m,n} z_{\gamma l}z_{k\beta}\frac{\partial}{\partial z_{\gamma\beta}}\Delta_{i_1,\dots,i_p}$$
$$= -\sum_{\alpha,\beta=1}^{p,p} z_{i_\alpha l}z_{k\beta}(-1)^{\alpha+\beta}\Delta^{1,\dots,\hat{\beta},\dots p}_{i_1,\dots,\hat{i}_\alpha,\dots,i_p}.$$

Now apply (5).

(ii) Use the formula of (i) in the case $i_1 = 1, \dots, i_p = p$. Suppose that $l \le p$. Then for all $l \ne \beta$

$$\Delta^{1,\dots,\beta-1,l,\beta+1,\dots,p}_{i_1,\dots,i_p} = 0,$$

since two columns of the determinant coincide. But if $k \le p$, $k \ne \alpha$, then

$$\Delta_{1,\dots,\alpha-1,k,\alpha+1,\dots,p} = 0$$

for the same reason. The formula follows.

(iii) Using (i) and formula (5), we write

$$I(i_1, \dots, i_p; q) = -\sum_{\alpha=1}^{p}(-1)^{p+\alpha}\Delta_{i_1,\dots,\hat{i}_\alpha,\dots,i_p}\sum_{\beta=1}^{q} z_{i_\alpha\beta}\Delta^{1,\dots,\beta-1,p,\beta+1,\dots,q}$$
$$= -\sum_{\beta=1}^{q}\Delta^{1,\dots,\beta-1,p,\beta+1,\dots,q}\Delta^{1,\dots,p-1,\beta}_{i_1,\dots,i_p}.$$

If $p > q$, the index β in the second sum is one of the numbers $1, \dots, p-1$, so all the summands vanish.

If $p \le q$, the determinant $\Delta^{1,\dots,\beta-1,p,\beta+1,\dots,q}$ vanishes for all $\beta \ne p$, and the only surviving term is $-\Delta^{1,\dots,q}\Delta_{i_1,\dots,i_p} = -\Delta_{i_1,\dots,i_p}\det_q$.

(iv) Apply (i) to the determinant $\Delta_{i_1,\dots,\hat{i}_s,\dots,i_p}$:

$$\begin{aligned} T_{i_s p}\left(\Delta_{i_1,\dots,\hat{i}_s,\dots,i_p}\right) &= -\sum_{\alpha=1,\alpha\neq s}^{p} z_{i_\alpha p}\Delta_{i_1,\dots,i_{\alpha-1},i_s,i_{\alpha+1},\dots,\hat{i}_s,\dots,i_p} \\ &= \sum_{\alpha=1}^{p} z_{i_\alpha p}(-1)^{\alpha+s}\Delta_{i_1,\dots,\hat{i}_\alpha,\dots,i_p} - z_{i_s p}\Delta_{i_1,\dots,\hat{i}_s,\dots,i_p}. \end{aligned}$$

Substituting this into the definition of $J(i_1,\dots,i_p)$ and using (5), we get

$$\begin{aligned} J(i_1,\dots,i_p) &= \sum_{\alpha,s=1}^{p,p}(-1)^{p+\alpha} z_{i_\alpha p}\Delta_{i_1,\dots,\hat{i}_\alpha,\dots,i_p} \\ &\quad - \sum_{s=1}^{p} z_{i_s,p}(-1)^{p+s}\Delta_{i_1,\dots,\hat{i}_s,\dots,i_p} \\ &= p\Delta_{i_1,\dots,i_p} - \Delta_{i_1,\dots,i_p} = (p-1)\Delta_{i_1,\dots,i_p}. \end{aligned}$$

We now introduce certain differential operators which will play a major role below:

$$\begin{aligned} d_{i_1,\dots,i_l} &= \Delta_{i_1,\dots,i_l}(T), & \overline{d}_{i_1,\dots,i_l} &= \Delta_{i_1,\dots,i_l}(\overline{T}), \\ d^*_{i_1,\dots,i_l} &= \Delta_{i_1,\dots,i_l}(T^*), & \overline{d}^*_{i_1,\dots,i_l} &= \Delta_{i_1,\dots,i_l}(\overline{T}^*). \end{aligned}$$

For convenience, we denote $d_{1,\dots,l}$ by d_l.

The following lemma defines the action of these operators on the highest weight vectors.

4.4.2. Lemma.

$$d_{i_1,\dots,i_l}\varphi_{\overline{k}}(Z) = (-1)^l\prod_{j=1}^{l}(k_j - j + 1)\Delta_{i_1,\dots,i_l}(Z)\varphi_{\overline{k}}(Z).$$

Proof. We use induction on l. For $l=1$ the formula holds, because

$$d_{i_1}\varphi_{\overline{k}} = T_{i_1 1}\varphi_{\overline{k}} = \sum_{s=1}^{m}\prod_{j=1,j\neq s}^{m}\det{}^{k_j-k_{j+1}}(k_s - k_{s+1})\det{}^{k_s-k_{s+1}} T_{i_1 1}(\det{}_s).$$

By Lemma 4.4.1, (ii), $T_{i_1 1}(\det_s) = -z_{i_1 1}\det_s$, and therefore

$$d_{i_1}\varphi_{\overline{k}} = -\sum_{s=1}^{m}(k_s - k_{s+1})z_{i_1 1}\varphi_{\overline{k}} = -k_1 z_{i_1 1}\varphi_{\overline{k}} = -k_1\Delta_{i_1}\varphi_{\overline{k}}.$$

Now suppose that the formula has been proved for $l-1$. Expanding the determinant $d_{i_1,\dots,i_l} = \Delta_{i_1,\dots,i_l}(T)$ with respect to the last column, we obtain

$$d_{i_1,\dots,i_l} = \sum_{\alpha=1}^{l}(-1)^{\alpha+l}T_{i_\alpha l}d_{i_1,\dots,\hat{i}_\alpha,\dots,i_l}.$$

By the induction hypothesis,

$$\begin{aligned} d_{i_1,\ldots,i_l}\varphi_{\overline{k}} &= (-1)^{l-1}\prod_{j=1}^{l-1}(k_j-j+1)\sum_{\alpha=1}^{l}(-1)^{\alpha+l}T_{i_\alpha l}\Delta_{i_1,\ldots,\hat{i}_\alpha,\ldots,i_l}\varphi_{\overline{k}} \\ &= (-1)^{l-1}\prod_{j=1}^{l-1}(k_j-j+1)\,[\Sigma_1+\Sigma_2]\,, \end{aligned}$$

where the sums

$$\Sigma_1 = \sum_{\alpha=1}^{l}(-1)^{\alpha+l}\varphi_{\overline{k}}T_{i_\alpha l}\Delta_{i_1,\ldots,\hat{i}_\alpha,\ldots,i_l} = \varphi_{\overline{k}}J(i_1,\ldots,i_l),$$

$$\Sigma_2 = \sum_{\alpha=1}^{l}(-1)^{\alpha+l}\Delta_{i_1,\ldots,\hat{i}_\alpha,\ldots,i_l}T_{i_\alpha l}\varphi_{\overline{k}}$$

are obtained as the result of the action of the first-order differential operator $T_{i_\alpha l}$ on the product $\Delta_{i_1,\ldots,\hat{i}_\alpha,\ldots,i_l}\varphi_{\overline{k}}$. The sum Σ_1 is evaluated by Lemma 4.4.1(iv):

$$\Sigma_1 = (l-1)\varphi_{\overline{k}}\Delta_{i_1,\ldots,i_l}.$$

As for Σ_2, we write, using the explicit expressions (2) for $T_{i_\alpha l}$ and $\varphi_{\overline{k}}$ and the definition of $I(i_1,\ldots,i_l;s)$:

$$\Sigma_2 = \sum_{s=1}^{m}(k_s-k_{s+1})\det{}_1^{k_1-k_2}\cdots\det{}_s^{k_s-k_{s+1}-1}\cdots\det{}_m^{k_m}\cdot I(i_1,\ldots,i_l;s),$$

and it follows from Lemma 4.4.1(iii) that

$$\Sigma_2 = -\sum_{s=l}^{m}(k_s-k_{s+1})\varphi_{\overline{k}}\Delta_{i_1,\ldots,i_l} = -k_l\varphi_{\overline{k}}\Delta_{i_1,\ldots,i_l}.$$

Hence $\Sigma_1+\Sigma_2 = -(k_k-l+1)\Delta_{i_1,\ldots,i_l}\varphi_{\overline{k}}$, and this completes the proof.

4.4.3. INCREASING SIGNATURES. PROPOSITION.

$$d_l\varphi_{\overline{k}} = (-1)^l\prod_{j=1}^{l}(k_j-j+1)\varphi_{\overline{k}+\epsilon_l}, \tag{i}$$

where $\epsilon_l = (\underbrace{1,\ldots,1}_{l},0,\ldots,0)$.

$$\varphi_{\overline{r}}(T)\varphi_{\overline{k}} = c(\overline{r},\overline{k})\varphi_{\overline{k}+\overline{r}}, \qquad c(\overline{r},\overline{k}) = \text{const}, \tag{ii}$$

where $c(\overline{r},\overline{k}) = 0$ *if and only if there exist numbers* t, j, $1\le t\le m$, $1\le j\le t$, *such that* $r_{t+1}-r_t+1\le k_j-j+1\le 0$. *If* $k_{\alpha+1}=\alpha$ *for some* α, $0\le\alpha\le m-1$, *then* $c(\overline{r},\overline{k})\ne 0$ *for signatures* $\overline{r}=(r_1,\ldots,r_{\alpha+1},0,\ldots,0)$, *and such signatures only.*

If $k_{\alpha+1} > \alpha$, *then* $c(\overline{r}, \overline{k}) \neq 0$ *for all* $\overline{r} = (r_1, \dots, r_{\alpha+1}, 0, \dots, 0)$.

PROOF. (i) Set $i_1 = 1, \dots, i_l = l$ in Lemma 4.4.2, and the formula follows.

(ii) Since

$$\varphi_{\overline{r}}(T) = \prod_{j=1}^{m} d_j^{r_j - r_{j+1}}, \qquad r_{m+1} = 0,$$

the repeated application of (i) yields

$$\varphi_{\overline{r}}(T)\varphi_{\overline{k}} = c\varphi_{\overline{k} + \sum_{j=1}^{m}(r_j - r_{j+1})\epsilon_j} = c\varphi_{\overline{k}+\overline{r}},$$

where

$$c = c(\overline{k}, \overline{r}) = \pm \prod_{t=1}^{m} \prod_{M=0}^{r_t - r_{t+1} - 1} \prod_{j=1}^{t} (k_j - j + 1 + M).$$

The first assertion about $c(\overline{k}, \overline{r})$ now follows.

Now suppose that $k_{\alpha+1} = \alpha$. Then $c \neq 0$ only if the product contains no factors with $t > \alpha$. This is true only provided $r_t - r_{t+1} = 0$ for all α, $\alpha < t \leq m$, i.e., if $r_{\alpha+1} = \cdots = r_m = r_{m+1} = 0$. Finally, if $k_{\alpha+1} > \alpha$, $r_{\alpha+2} = \cdots = r_m = 0$, then none of the factors in $c(\overline{r}, \overline{k})$ vanish.

We will now describe the action of operators $\overline{d}^*_{i_1, \dots, i_l}$ on the highest weight vectors.

4.4.4. LEMMA.

(i) $\overline{d}^*_{i_1, \dots, i_l}\varphi_{\overline{k}} = \prod_{j=1}^{l}(k_j - j + 1 + n)\Delta_{i_1, \dots, i_l}\varphi_{\overline{k}}$,

(ii) $\overline{d}^*_l \varphi_{\overline{k}} = \prod_{j=1}^{l}(k_j - j + 1 + n)\varphi_{\overline{k}+\epsilon_l}$.

PROOF. (i) We use induction on l. Application of Propositions 4.2.1 and 4.4.3 yields

$$\overline{d}^*_{i_1}\varphi_{\overline{k}} = \left(-d_{i_1} + nz_{i_1 1}\right)\varphi_{\overline{k}} = (k_1 + n)\varphi_{\overline{k}}.$$

Suppose the formula is true for $l - 1$. Expand the determinant $\overline{d}^*_{i_1, \dots, i_l}$ with respect to the last column:

$$\overline{d}^*_{i_1, \dots, i_l} = \sum_{\alpha=1}^{l}(-1)^{\alpha+l}\overline{T}^*_{i_\alpha, l}\overline{d}^*_{i_1 \cdots \hat{i}_\alpha \cdots i_l}.$$

By the induction hypothesis,

$$\begin{aligned}\overline{d}^*_{i_1,\dots,i_l}\varphi_{\overline{k}} &= \prod_{j=1}^{l-1}(k_j - j + 1 + n)\sum_{\alpha=1}^{l}(-1)^{\alpha+l}\overline{T}^*_{i_\alpha,l}\left(\Delta_{i_1,\dots,\hat{i}_\alpha,\dots,i_l}\varphi_{\overline{k}}\right)\\ &= \prod_{j=1}^{l-1}(k_j - j + 1 + n)\left[-\sum_{\alpha=1}^{l}(-1)^{\alpha+l}T_{i_\alpha,l}\left(\Delta_{i_1,\dots,\hat{i}_\alpha,\dots,i_l}\varphi_{\overline{k}}\right)\right.\\ &\qquad\left.+n\sum_{\alpha=1}^{l}(-1)^{\alpha+l}z_{i_\alpha l}\Delta_{i_1,\dots,\hat{i}_\alpha,\dots,i_l}\varphi_{\overline{k}}\right].\end{aligned}$$

The first sum in the brackets equals $\Sigma_1 + \Sigma_2$, as in the proof of Lemma 4.4.2; hence, it is equal to $(-k_l + l - 1)\Delta_{i_1,\dots,i_l}\varphi_{\overline{k}}$. By (5), the second sum equals $n\Delta_{i_1,\dots,i_l}\varphi_{\overline{k}}$. Substituting this into (*), we get (i). Formula (ii) is the particular case of (i) for $i_1 = 1, \dots, i_l = l$.

The following corollary of the above lemma will play an important role in the proof of the main theorem.

4.4.5. PROPOSITION (Decreasing signatures). *Let* $k_1 \geq k_2 \geq \dots \geq k_q > k_{q+1} \geq \dots \geq k_m$. *Then*

$$\left\langle \overline{d}_q\varphi_{\overline{k}}, \varphi_{\overline{k}-\epsilon_q}\right\rangle \neq 0.$$

PROOF. It is easy to see that $\overline{k} - \epsilon_q$ is a signature. By Lemma 4.4.4(ii),

$$\left\langle \overline{d}_q\varphi_{\overline{k}}, \varphi_{\overline{k}-\epsilon_q}\right\rangle = \left\langle \varphi_{\overline{k}}, \overline{d}^*_q\varphi_{\overline{k}-\epsilon_q}\right\rangle = \prod_{j=1}^{q}(k_j - j + 1 + n)\langle\varphi_{\overline{k}}, \varphi_{\overline{k}}\rangle \neq 0.$$

4.5. PROOF OF THEOREM 4.1.1. Note first of all that it suffices to consider the case $X = H^2(St_{m,n})$. This follows from the fact that, for any closed K-invariant subspace of $C(St_{m,n})$ or $L^p(St_{m,n})$,

$$W = \mathrm{cl}\left[W \cap L^2(St_{m,n}\right].$$

This in turn follows in the standard way from the invariance of W with respect to convolutions by the functions on K (see [26] for a proof in the case $m = 1$).

Thus, we shall consider only subspaces W of $H^2(St_{m,n})$. If W is invariant with respect to the group $U(m, n)$, then the same is true for the smaller group K. Therefore, we can express W as a direct sum of irreducible subspaces $H_{\overline{k}}$:

$$W = \mathrm{cl}_{L^2}\left[\bigoplus_{\overline{k}\in\Omega} H_{\overline{k}}\right], \qquad \Omega \subset (\mathbb{Z}_+)^m,$$

where Ω is the set of signatures $\overline{k}$ such that the orthogonal projections

$$P_{\overline{k}}: L^2(St_{m,n}) \to H_{\overline{k}}$$

do not vanish identically on W, that is,

$$\langle f, \varphi_{\overline{k}}\rangle \neq 0$$

for some function $f \in W$. The space W defined by Ω will be denoted by $(H^2)_\Omega$. We are going to determine when this space is invariant under $U(m, n)$.

It follows from Propositions 4.4.3 and 4.4.5 that if $(H^2)_\Omega$ is invariant under $U(m, n)$, then:

(a) if $\overline{k} = (k_1, \dots, k_m) \in \Omega$ and $k_j \neq j-1$ for all $j = 1, \dots, l$, then

$$\sigma_l^+ \overline{k} = \overline{k} + \epsilon_l = (k_1 + 1, \dots, k_l + 1, k_{l+1}, \dots, k_m) \in \Omega;$$

(b) if $\overline{k} \in \Omega$ and $k_l > k_{l+1}$, then

$$\sigma_l^- \overline{k} = \overline{k} - \epsilon_l = (k_1 - 1, \dots, k_l - 1, k_{l+1}, \dots, k_m) \in \Omega.$$

In other words, $\sigma_l^- \overline{k} \in \Omega$ if the sequence $\overline{k} - \epsilon_l$ is a signature, i.e., if the condition that the components are decreasing and nonnegative is fulfilled.

The only sets Ω invariant under the operations $\sigma_l^{\pm}$, $l = 1, \dots, m$, are the sets

$$\Omega_q = \left\{\overline{k} = (k_1, \dots, k_m) \in \mathbb{Z}_+^m : k_1 \geq \cdots \geq k_m \geq 0,\ k_{m+1} = 0,\ k_{q+1} \leq q\right\},$$

$q = 0, 1, \dots, k$. This proves the following statement.

4.5.1. PROPOSITION. *A closed $U(m, n)$-invariant subspace of $H^2(St_{m,n})$ is one of the following:*

$$(H^2)_{\Omega_q}, \quad 0 \leq q \leq m; \qquad (H^2)_{\Omega_m} = H^2(St_{m,n}).$$

Our description of the invariant subspaces of $H^2(St_{m,n})$ will be complete if we show that every subspace $(H^2)_{\Omega_q}$ is invariant under the action of the group $U(m, n)$.

4.5.2. PROPOSITION. $(H^2)_{\Omega_q}$ *coincides with the space V_s defined in* §4.1, *i.e., the closed $U(m, n)$-cyclic span $\mathrm{cl}_{L^2}\left[\det_m^q\right]_{U(m,n)}$ of the function $\det_m^q$.*

We first prove the following lemma.

4.5.3. LEMMA. *Let $m = n$ and $0 \leq q < q' \leq n$. Then one has*

$$\left\langle T_\omega \det_n^q, \det_n^{q'} \right\rangle = 0$$

for any $\omega \in U(n, m)$.

PROOF. By the Taylor formula, it suffices to prove the equality

$$\left\langle \mathscr{D} \det_n^q, \det_n^{q'} \right\rangle = 0$$

for an arbitrary differential operator $\mathscr{D}$ in the universal enveloping algebra $\mathfrak{A}$ of the Lie algebra $\mathfrak{g}$. Since $\mathfrak{A}$ consists of the polynomials in the operators L_{ij}, R_{ij}, T_{ij}, $\overline{T}_{ij}$, we may assume without loss of generality that $\mathscr{D}$ is a monomial in these operators. Using the commutation relations

$$\left[T_{ks}, \overline{T}_{ij}\right] = \delta_{sj}L_{ki} + \delta_{ki}R_{sj}, \qquad \left[L_{ks}, L_{ij}\right] = \delta_{is}L_{kj} - \delta_{kj}L_{is},$$
$$\left[L_{ks}, T_{ij}\right] = \delta_{si}T_{kj}, \qquad \left[L_{ks}, \overline{T}_{ij}\right] = -\delta_{ki}T_{sj},$$
$$\left[R_{ks}, R_{ij}\right] = \delta_{si}R_{kj} - \delta_{kj}R_{is}, \qquad \left[R_{ks}, T_{ij}\right] = \delta_{sj}T_{ki},$$
$$\left[R_{ks}, \overline{T}_{ij}\right] = -\delta_{kj}\overline{T}_{is}, \qquad \left[L_{ks}, R_{ij}\right] = \left[T_{ks}, T_{ij}\right] = \left[\overline{T}_{ks}, \overline{T}_{ij}\right] = 0,$$

we represent $\mathscr{D}$ as a linear combination of ordered monomials of the form

$$\overline{T}^{\alpha}T^{\beta}X^{\gamma}, \qquad \overline{T}^{\alpha} = \prod \overline{T}_{ij}^{\alpha_{ij}}, \quad T^{\beta} = \prod T_{ij}^{\beta_{ij}},$$

where $X^{\gamma} \in \mathfrak{k}^{\mathbb{C}}$ is a monomial in L_{ij}, R_{ij}.

It follows from the equalities

$$L_{ij}\det_n = R_{ij}\det_n = \delta_{ij}\det_n$$

that $\det_n^q$ is an eigenfunction of the operators in $\mathfrak{k}^{\mathbb{C}}$. Therefore, the operator X^{γ} may be omitted, and it suffices to prove that the scalar product

$$\left\langle \overline{T}^{\alpha}T^{\beta}\det_n^q, \det_n^{q'} \right\rangle = \left\langle T^{\beta}\det_n^q, \left(\overline{T}^{\alpha}\right)^* \det_n^{q'} \right\rangle$$

vanishes. By Lemma 4.4.1(ii) and Proposition 4.2.1,

$$\overline{T}_{ij}^*\det_n^{q'} = \left(-T_{ij} + nz_{ij}\right)\det_n^{q'} = (-q' + n)z_{ij}\det_n^{q'},$$

so that

$$\left(\overline{T}^{\alpha}\right)^*\det_n^{q'} = Q(Z)\det_n^{q'},$$

where Q is a polynomial in Z.

On the other hand, by (4) we can represent T^{β} as a linear combination of multiple commutators

$$[X, \varphi_{\bar{r}}(T)]^{(j)}, \qquad X \in \mathfrak{k}^{\mathbb{C}},$$

so it is enough to verify that

$$\left\langle [X, \varphi_{\bar{r}}(T)]^{(j)}\det_n^q, Q(Z)\det_n^{q'} \right\rangle = 0.$$

One more simplifying remark: since $\det_n^q$ is an eigenfunction of $X \in \mathfrak{k}^{\mathbb{C}}$, the multiple commutator

$$[X, \varphi_{\bar{r}}(T)]^{(j)}\det_n^q$$

is a linear combination of expressions of the form $X^s \varphi_{\bar{r}}(T)\det_n^q$. Since

$$\left\langle X^s \varphi_{\bar{r}}(T)\det_n^q, Q(Z)\det_n^{q'}\right\rangle = \left\langle \varphi_{\bar{r}}(T)\det_n^q, (X^*)^s Q(Z)\det_n^{q'}\right\rangle$$
$$= \left\langle \varphi_{\bar{r}}(T)\det_n^q, Q_1(Z)\det_n^{q'}\right\rangle,$$

where Q_1 is a polynomial, we are left with the task of proving the equality

$$\left\langle \varphi_{\bar{r}}(T)\det_n^q, Q_1(Z)\det_n^{q'}\right\rangle = 0.$$

By Lemma 4.4.3(ii),

$$\varphi_{\bar{r}}(T)\det_n^q = c\varphi_{\bar{q}+\bar{r}}, \qquad \bar{q} = (q, \ldots, q)$$

and $c = c(\bar{r}, \bar{q}) = 0$ if $r_{q+1} > 0$. If $r_{q+1} = 0$, then $r_{q+2} = \cdots = r_n = 0$ and therefore

$$\left\langle \varphi_{\bar{r}}(T)\det_n^q, Q_1(Z)\det_n^{q'}\right\rangle = c(\bar{r}, \bar{q})\left\langle \varphi_{\bar{q}+\bar{r}}, Q_1(Z)\det_n^{q'}\right\rangle$$
$$= c(\bar{r}, \bar{q})\varphi_{\bar{q}+\bar{r}} \int_{St_{n,n}} \det_1^{r_1-r_2}(Z)\cdots\det_q^{r_q}(Z)\,\overline{Q_1(Z)}\overline{\det_n^{q'-q}(Z)}\,d\sigma(Z).$$

Since $q < n$, the holomorphic factors in the integrand do not depend on the elements of the last column of Z. Change variables by

$$(Z^1, \ldots, Z^n) \to (Z^1, \ldots, Z^{n-1}, e^{i\varphi}Z^n)$$

and integrate both sides of the equality with respect to φ over $[0, 2\pi]$. Since the integral is invariant under this change of variables, this yields

$$2\pi\langle \varphi_{\bar{r}}(T)\det_n^q, Q_1(Z)\det_n^{q'}\rangle = c(\bar{r}, \bar{q})\langle \varphi_{\bar{q}+\bar{r}}, Q_1(Z)\det_n^{q'}\rangle$$
$$= c(\bar{r}, \bar{q}) \int_{St_{n,n}} \det_1^{r_1-r_2}(Z)\cdots\det_q^{r_q}(Z)\overline{Q}_1(Z^1, \ldots, Z^{n-1}, 0)$$
$$\times\, \overline{\det_n^{q'-q}(Z^1, \ldots, Z^{n-1}, 0)}\,d\sigma(Z) = 0.$$

PROOF OF PROPOSITION 4.5.2. By Proposition 4.5.1, the space V_q is $(H^2)_{\Omega_{q'}}$ for some q', $0 \le q' \le m$, and it suffices to prove that $q = q'$. Since the function $\det_m^q = \varphi_{(q,\ldots,q)}$ belongs to $(H^2)_{\Omega^{q'}}$, it follows that $(q, \ldots, q) \in \Omega_{q'}$, and, by the definition of the latter, $q \le q'$.

In order to prove the reverse inequality, we distinguish two cases.

(1) $m = n$. Since $\det_n^{q'} \in (H^2)_{\Omega^{q'}}$ and $(H^2)_{\Omega^{q'}} = V_q$, it follows from Lemma 4.5.3 that the inequality $q < q'$ cannot hold.

(2) $m < n$. There is a natural embedding of $U(m, n)$ into $U(n, n)$:

$$\omega = \begin{pmatrix} A & B \\ C & D \end{pmatrix} \to \tilde{\omega} = \begin{pmatrix} A & 0 & B \\ 0 & 1_{n-m} & 0 \\ C & 0 & D \end{pmatrix},$$

where A, B, C, D are matrices of order $m \times m$, $m \times n$, $n \times m$, $n \times n$ respectively. The manifold $St_{m,n}$ is a quotient space of $St_{n,n} = U(n)$:

$$St_{m,n} = St_{n,n}/U(n-m).$$

Denote the natural projection $St_{n,n} \to St_{m,n}$, by σ, which carries an $n \times n$ matrix Z onto its first m rows. Let σ^* denote the induced map of function spaces: $\sigma^* f = f \circ \sigma$.

Since the function $\det_m^{q'}$ belongs to the space $(H^2)_{\Omega_{q'}} = V_q$, its pullback $\sigma^* \det_m^{q'}$ belongs to the closure of the $U(n, n)$-cyclic span of the function

$$\sigma^* \det_m^q = \det_m^q = \varphi_{(\underbrace{q, \dots, q}_{m}, 0, \dots, 0)}$$

by definition of V_q, since the automorphisms in $U(m, n)$ can be extended to automorphisms in $U(n, n)$. But we have already dealt with the case $m = n$, so the spaces $\mathrm{cl}_{L^2}[\det_n^q]_{U(n,n)}$ and $(H^2)_{\Omega_q}$ coincide. Therefore, the function $\sigma^* \det_m^{q'}$, as an element of the closed cyclic span of $\sigma^* \det_m^q$, also belongs to $(H^2)_{\Omega_q}$. Thus the signature $(q' \dots, q', 0, \dots, 0)$ belongs to Ω_q, and this implies, by the definition of the latter, that $q' \le q$. This proves Proposition 4.5.2.

The theorem formulated at the beginning of this section follows from Propositions 4.5.1 and 4.5.2 in the case $X = H^2(St_{m,n})$. By a previous remark, the theorem is also true for $X = A(St_{m,n})$ and $X = L^p(St_{m,n})$.

§5. Möbius subspaces of $C(\overline{D})$

We shall obtain partial characterizations of closed Möbius subspaces and subalgebras of $C(\overline{D})$ on bounded symmetric domains. A complete classification of closed Möbius subalgebras for a ball was given by Nagel and Rudin [84]. Subspaces (even in the case of the ball) have received much less attention.

Let $\overline{D}$ denote the closure of a domain D. Let $C_0(D)$ denote the algebra of functions in $C(\overline{D})$ that vanish on ∂D. Let $A(D)$, $\overline{A}(D)$, $H(D)$ denote the subspaces of $C(\overline{D})$ consisting respectively of holomorphic, antiholomorphic, and Aut(D)-harmonic functions. A function is Aut(D)-harmonic if it belongs to the kernel of the Aut(D)-invariant Laplace operator.

5.1. Theorem. *Let D be an irreducible bounded symmetric domain. Then*

(1) *every closed Möbius subspace $Y \subset C(\overline{D})$ such that $Y \cap C_0(D) = \{0\}$ is contained in $H(D)$;*

(2) *every closed Möbius subalgebra $A \subset C(\overline{D})$ such that $A \cap C_0(D) = \{0\}$ is one of the algebras $\{0\}$, $\mathbb{C}$, $A(D)$, $\overline{A}(D)$.*

PROOF. Part (2) follows from (1) and Theorem 3.2, Chapter 3, because, by the maximum principle, a subspace of $H(D)$ is completely determined by its restrictions to the Shilov boundary.

To prove (1) we will show that Y is the space of Aut(D)-invariant Poisson integrals of the restrictions $f|_{\partial D}$ of its elements to ∂D. Let $f \in Y$ and $w \in D$. Consider the average

$$S_w f = \int_{\mathrm{Aut}(D,w)} (T_k f)\, dk$$

of f over the stationary subgroup of w. Since $S_w f$ is constant on ∂D and belongs to Y, it follows that $S_w f - (S_w f)(z_0) \in Y \cap C_0(D)$ and hence $(S_w f)(z) = (S_w f)(z_0)$, where z_0 is an arbitrary fixed point in ∂D, $z \in D$. Setting $z = w$ and taking into account that

$$(S_w f)(z_0) = P\left(f|_{\partial D}\right)(w)$$

(see formula (1), §2), we get

$$f(w) = (S_w f)(w) = P\left(f|_{\partial D}\right)(w).$$

The next theorem follows from Theorem 3.2, Chapter III.

5.2. THEOREM. *Let D be an irreducible bounded symmetric domain. Then every closed Möbius subalgebra $B \subset C(\overline{D})$ such that $C_0(D) \subset B$ can be represented as*

$$B = C_0(D) + A,$$

where A is one of the algebras $\{0\}$, $\mathbb{C}$, $A(D)$, $\overline{A}(D)$, $C(\overline{D})$.

§6. Möbius-invariant spaces in the complex ball. Separation of the $\mathscr{M}$-harmonic component

6.1. Denote by $B = \{z \in C^n : |z|^2 = |z_1|^2 + \cdots + |z_n|^2 < 1\}$ the unit ball in the space $\mathbb{C}^n$, by $\overline{B}$ and S the closure and the boundary of B, respectively, by $C(\overline{B})$ the space of continuous complex functions on B with the sup-norm, and by $C_0(B)$ the subspace of the functions that are equal to zero on S.

Let $\mathscr{M}$ be the group of complex linear-fractional automorphisms of the ball B (the Möbius group), and let $d\tau = (1-|z|^2)^{-2n} d\sigma_B$ be the $\mathscr{M}$-invariant volume, where σ_B is the Lebesgue measure on B.

A space of functions Y on $\overline{B}$ is said to be $\mathscr{M}$-invariant if $f \in Y$, $\varphi \in \mathscr{M}$, implies $f \circ \varphi \in Y$. An example of such a space in $C(\overline{B})$ is provided by the space $H(B)$ consisting of the $\mathscr{M}$-harmonic functions, i.e., the solutions of the equation

$$\widetilde{\Delta} u = (1 - |z|^2) \sum_{i,j=1}^{n} (\delta_{ij} - z_i \overline{z}_j) \frac{\partial^2 u}{\partial z_i \partial \overline{z}_j} = 0.$$

The $\mathscr{M}$-invariance of the space $H(B)$ is a consequence of the invariance of the operator $\widetilde{\Delta}$ under the action of the group $\mathscr{M}$. The space $H(B)$ has closed $\mathscr{M}$-invariant subspaces: the spaces $A(B)$, $\overline{A(B)}$, and $\mathrm{Ph}(B)$ of holomorphic, anti-holomorphic, and pluriharmonic functions, respectively, and also $\{0\}$ and the space $\mathbb{C}$ of constant functions. These spaces exhaust the list of closed $\mathscr{M}$-invariant subspaces of $C(\overline{B})$ with the property that their intersection with $C_0(B)$ is equal to zero [84].

A different and considerably less transparent class of $\mathscr{M}$-invariant subspaces is that of subspaces of $C_0(B)$. There arises the question whether all closed $\mathscr{M}$-invariant subspaces in $C(\overline{B})$ can be constructed from invariant subspaces of the space $C_0(B)$ and the spaces of $\mathscr{M}$-harmonic functions listed above.

In this connection Nagel and Rudin [84] (see also [26, 13.3.4]) conjectured that any $\mathscr{M}$-invariant subspace $Y \subset C(\overline{B})$ admits a direct sum decomposition

$$Y = Y_0 \oplus F, \tag{1}$$

where Y_0 and F are closed $\mathscr{M}$-invariant subspaces of $C_0(B)$ and $H(B)$, respectively. Since all spaces F are known, the existence of such a decomposition would allow one to disregard boundary values and thereby reduce the problem of describing the $\mathscr{M}$-invariant spaces over the closed ball to the case of spaces with null boundary values.

The problem of the existence of a decomposition (1) is equivalent to the following problem. Consider the Poisson operator P corresponding to the $\mathscr{M}$-invariant Laplace operator $\widetilde{\Delta}$:

$$(Pf)(z) = \int_S f(\zeta)\frac{(1-|z|^2)^n}{|1-\langle z, \zeta\rangle|^{2n}}\, d\sigma_S(\zeta),$$

where $\langle z, \zeta\rangle = \sum_{i=1}^n z_i\overline{\zeta}_i$, and σ_S is the Lebesgue measure on S. Let $j : C(\overline{B}) \to C(S)$ denote the operator of restriction to S. One is asking whether the operator $P \circ j$ maps every closed $\mathscr{M}$-invariant subspace $Y \subset C(\overline{B})$ into itself. In other words, is it true that any $\mathscr{M}$-harmonic function that coincides on S with a function belonging to Y does itself belong to Y? If it is so, decomposition (1) is obvious: $f = [f - (P\circ j)f] + (P\circ j)f$.

In this section, we give a positive answer to this question (and thus establish the validity of decomposition (1)) for a certain class of $\mathscr{M}$-invariant subspaces of $C(\overline{B})$.

Let us formulate our result. We denote by W the class of functions whose $\mathscr{M}$-invariant Laplacian is square-integrable with respect to the $\mathscr{M}$-invariant volume:

$$W = \left\{f \in C(\overline{B}) \cap C^2(B) : \int_B |\widetilde{\Delta} f(z)|^2 d\tau(z) < \infty\right\}.$$

Theorem. *Let Y be a closed $\mathscr{M}$-invariant subspace of $C(\overline{B})$. Suppose that $Y \cap W$ is uniformly dense in Y. Then the operator $P \circ j$ maps Y into itself and decomposition* (1) *holds.*

6.2. The Operator $\widetilde{\Delta}$. Let us study the action of the $\mathscr{M}$-invariant Laplace operator on Möbius-invariant spaces.

Consider the following one-parameter subgroups of the group $\mathscr{M}$:

$$\varphi_t^k(z) = \frac{(t + z_k)e_k + \sqrt{1-t^2}(z - z_k e_k)}{1 + tz_k},$$

$$\psi_t^k(z) = \frac{(it + z_k)e_k + \sqrt{1-t^2}(z - z_k e_k)}{1 - itz_k},$$

where e_k are the standard unit vectors in C^n, $k = 1, \dots, n$. Let X_k and Y_k denote the infinitesimal operators of the subgroups:

$$X_k f = \frac{d}{dt}(f \circ \varphi_t^k)|_{t=0}, \qquad Y_k f = \frac{d}{dt}(f \circ \psi_t^k)|_{t=0},$$

and set $Q_k = (1/2)\,(X_k - iY_k)$, $\overline{Q}_k = (1/2)\,(X_k + iY_k)$.

In complex coordinates the operators Q_k and $\overline{Q}_k$ have the expressions

$$Q_k = D_k - \overline{z}_k \sum_{i=1}^n \overline{z}_i \overline{D}_i, \qquad \overline{Q}_k = \overline{D}_k - z_k \sum_{i=1}^n z_i D_i,$$

$$D_k = \frac{\partial}{\partial z_k}, \quad \overline{D}_k = \frac{\partial}{\partial \overline{z}_k}.$$

Now set

$$\begin{aligned} R &= \frac{1}{2}\sum_{k=1}^n [Q_k\overline{Q}_k + \overline{Q}_k Q_k] \\ &= \sum_{k=1}^n D_k\overline{D}_k - \sum_{i,j=1}^n z_i z_j D_i D_j - \sum_{i,j=1}^n \overline{z}_i\overline{z}_j \overline{D}_i\overline{D}_j \\ &\quad + |z|^2 \sum_{i,j=1}^n z_i\overline{z}_j D_i\overline{D}_j - \frac{n+1}{2}\sum_{i=1}^n z_i D_i - \frac{n+1}{2}\sum_{i=1}^n \overline{z}_i\overline{D}_i. \end{aligned} \tag{2}$$

Next, consider the following basis in the Lie algebra $\mathfrak{U}(n)$ of the isotropy subgroup of zero, $U(n) \subset \mathscr{M}$:

$$A_{kl}^+ = i(E_{kl} - E_{lk}), \qquad A_{kl}^- = E_{kl} - E_{lk},$$

where $(E_{kl})_{ij} = \delta_{ik}\delta_{lj}$. The operators of the corresponding derived (infinitesimal) representation

$$D\left(A_{kl}^\pm\right) = \frac{d}{dt}\left(f \circ \exp\left(tA_{kl}^\pm\right)\right)|_{t=0}$$

have the complex-coordinate expressions

$$D\left(A_{kl}^\pm\right) = \varepsilon^\pm\left[z_l D_k \pm z_k D_l \mp \overline{z}_l\overline{D}_k - \overline{z}_k\overline{D}_l\right],$$

where $\varepsilon^+ = i$, $\varepsilon^- = 1$.

Introducing the operator

$$V=-\frac{1}{8}\sum_{k,l=1}^{n}\left[D\left(A_{kl}^{+}\right)\right]^{2}-\frac{1}{8}\sum_{k,l=1}^{n}\left[D(A_{kl}^{-})\right]^{2}+\frac{1}{4}\left[\sum_{k=1}^{n}D\left(A_{kk}^{+}\right)\right]^{2},$$

we obtain

$$\begin{aligned} V=&-|z|^{2}\sum_{k=1}^{n}D_{k}\overline{D}_{k}+\sum_{i,j=1}^{n}z_{i}z_{j}D_{i}D_{j}+\sum_{i,j=1}^{n}\overline{z}_{i}\overline{z}_{j}\overline{D}_{i}\overline{D}_{j}\\ &-\sum_{i,j=1}^{n}z_{i}\overline{z}_{j}D_{i}\overline{D}_{j}+\frac{n+1}{2}\sum_{i=1}^{n}z_{i}D_{i}+\frac{n+1}{2}\sum_{i=1}^{n}\overline{z}_{i}\overline{D}_{i}. \end{aligned} \tag{3}$$

Adding expression (2) and (3) we arrive at the following expression for the operator $\widetilde{\Delta}$:

$$\begin{aligned} \widetilde{\Delta}&=\sum_{k=1}^{n}D_{k}\overline{D}_{k}+|z|^{2}\sum_{i,j=1}^{n}z_{i}\overline{z}_{j}D_{i}\overline{D}_{j}-|z|^{2}\sum_{k=1}^{n}D_{k}\overline{D}_{k}-\sum_{i,j=1}^{n}z_{i}\overline{z}_{j}D_{i}\overline{D}_{j}\\ &=R+V=\frac{1}{2}\sum_{k=1}^{n}\left[Q_{k}\overline{Q}_{k}+\overline{Q}_{k}Q_{k}\right]-\frac{1}{8}\sum_{k,l=1}^{n}\left\{\left[D(A_{kl}^{+})\right]^{2}\right\}+D\left(A_{kl}^{-}\right)^{2}. \end{aligned} \tag{4}$$

LEMMA. *Let Y be a closed $\mathscr{M}$-invariant subspace of $C(\overline{B})$. Then there exists a dense subspace $Y^{\omega}\subset Y\cap C^{\infty}(B)$ that is mapped by $\widetilde{\Delta}$ into itself.*

PROOF. Consider the representation $T:\varphi\to T_{\varphi}$, $\varphi\in\mathscr{M}$, of the semisimple group $\mathscr{M}$ in the Banach space Y by means of the translation operators $T_{\varphi}f=f\circ\varphi^{-1}$, $f\in Y$. Then for Y^{ω} one can take the space of analytic vectors of the representation T. The space Y^{ω} is dense in Y and invariant under the operators of the derived representation of the Lie algebra of $\mathscr{M}$ [24]. This implies the required assertion, because in view of (4) $\widetilde{\Delta}$ is expressible through the operators of the derived representation.

6.3. Spherical-Mean Operators. Consider the spherical-mean operators

$$(S_{w}f)(z)=\int_{U(n)}f(\varphi(uw))\,du,$$

where for φ one can take any automorphism satisfying the condition $\varphi(0)=z$. Such an automorphism can explicitly be given by the formula [84]

$$\varphi_{z}(w)=\frac{z-w-P_{z}w-\sqrt{1-|z|^{2}}(z-P_{z}w)}{1-\langle w,z\rangle},\qquad P_{z}(w)=\frac{\langle w,z\rangle}{|z|^{2}}z.$$

LEMMA. *Let Z be a closed $\mathscr{M}$-invariant subspace of the space $C_{0}(B)$. Set $E=Z\cap L^{2}(B,\tau)$, and suppose that E is uniformly dense in Z. Then $S_{w}:Z\to X$ for any $\omega\in B$.*

PROOF. Pick an $f \in Z$ and assume first that f is a radial function, i.e., $f(z) = f(|z|)$. From the equality $|\varphi_z(uw)| = |Q_w(uz)|$, $u \in U(n)$ (see [69, 2.2.2]) we obtain

$$(S_w f)(z) = \int_{U(n)} f(\varphi_w(uz))\, du\,,$$

which in view of the invariance of the space Z implies that $S_w f \in Z$.

The same holds true in the case in which f is a linear combination of translates of spherical functions in Z, because the operators S_w commute with the translation operators T_φ, $\varphi \in \mathcal{M}$. Therefore, the lemma will be proved if one verifies that the set Z_{rad} of such linear combinations is uniformly dense in Z. To this end it suffices to prove the inclusion $Z_{\mathrm{rad}}^\perp \subset Z^\perp$ of annihilators in the dual of the space $C_0(B)$.

Pick a measure $\mu \in Z_{\mathrm{rad}}^\perp$ and consider its convolution

$$\mu_\psi = \int_{\mathcal{M}} (\mu \circ \psi) d\psi$$

with an arbitrary function $\psi \in C(\mathcal{M})$ with compact support. Here $d\psi$ is the Haar measure on $\mathcal{M}$. Clearly, $\mu_\psi \in Z_{\mathrm{rad}}^\perp$ and the measure μ_ψ is absoutely continuous with repect to the measure τ. As a matter of fact, $\mu_\psi = h\tau$, where $h \in L^2(B\,,\tau) \cap C(B)$.

Let h_1 be the orthogonal projection of h on the closure of the space E in $L^2(B\,,\tau)$. Average the function h_1 over the group $U(n)$:

$$h_1^{\mathrm{rad}} = \int_{U(n)} (h_1 \circ u)\, du.$$

Then h_1^{rad} belongs to the closure of the space $Z_{\mathrm{rad}} \cap L^2(B\,,\tau)$ in $L^2(b\,,\tau)$. Since $\mu_\psi \in Z_{\mathrm{rad}}^\perp$, one has that $h \perp Z_{\mathrm{rad}}$; in particular, $h \perp h_1^{\mathrm{rad}}$. Consequently,

$$\int_B |h_1^{\mathrm{rad}}(z)|^2 d\tau(z) = \int_B h_1(z)\overline{h_1^{\mathrm{rad}}(z)}\, d\tau(z) = \int_B h(z)\overline{h_1^{\mathrm{rad}}(z)}\, d\tau(z) = 0\,.$$

This implies $h_1(0) = h_1^{\mathrm{rad}}(0) = 0$. Replacing h_1 by $h_1 \circ \varphi$ in this argument, one obtains $h_1(\varphi(0) = 0$ for all $\varphi \in \mathcal{M}$. Therefore, $h_1 = 0$ and $h \perp E$. It follows that $\mu_\psi \in E^\perp$, and, since E is dense in Z, $\mu_\psi \in Z^\perp$. The measures μ_ψ approximate μ in the weak topology, so $\mu \in Z^\perp$ and the lemma is proved.

6.4. PROOF OF THE THEOREM. Suppose the space Y satisfies all conditions of the theorem, and let Y^ω be the space whose existence is asserted by Lemma 6.2.

LEMMA 1. *The space $Y^\omega \cap W$ is uniformly dense in Y.*

PROOF. Pick $f \in Y \cap W$. Denote by $\varphi(t, y)$ a solution of the heat equation on the group $\mathscr{M}$. Then the function

$$f_t = \int_{\mathscr{M}} \varphi(t, y^{-1}) T_y f dy$$

belongs to the space Y^{ω} of analytic vectors of the representation T [24]. Since $\varphi(t, \cdot) \in L^1(\mathscr{M})$ and $\widetilde{\Delta} f \in L^2(B, \tau)$, the function $\widetilde{\Delta} f_t$ belongs to $L^2(B, \tau)$, which, in view of the definition of the space W, means that $f_t \in W$. Further, since $f_t \to f$ in the uniform norm as $t \to 0$, one concludes that $Y^{\omega} \cap W$ is dense in the space $Y \cap W$. It remains to remark that the latter is dense in Y by the hypothesis of the theorem.

LEMMA 2. *Spherical-mean operators S_w, $w \in B$, map Y into itself.*

PROOF. By the preceding lemma, it suffices to show that $S_w f \in Y$ for all $f \in Y^{\omega} \cap W$.

Consider the range $\widetilde{\Delta}(Y^{\omega} \cap W)$ of the operator $\widetilde{\Delta}$. By Lemma 6.2 and the definition of the space W, $\widetilde{\Delta}(Y^{\omega} \cap W) \subset Y \cap L^2(B, \tau)$. In particular, $\widetilde{\Delta}(Y^{\omega} \cap W) \subset C_0(B)$.

Let Z denote the uniform closure of $\widetilde{\Delta}(Y^{\omega} \cap W)$. Then the space Z satisfies all conditions of Lemma 3.1. If $f \in Y^{\omega} \cap W$, one has $\widetilde{\Delta} f \in Z$ and, by Lemma 6.3, $S_w(\widetilde{\Delta} f) \in Z$.

Fix $f \in Y^{\omega} \cap W$, and let μ be an arbitrary measure in the closed ball $\overline{B}$ that belongs to the annihilator $Y^{\perp}$ in the dual of the space $C(\overline{B})$. Set

$$a_{\mu, f}(w) = \int_{\overline{B}} (S_w f)(z)\, d\mu(z).$$

Since $S_w f = f$ on S, one can write

$$a_{\mu, f}(w) = \int_B (S_w f)(z)\, d\mu + \int_S f\, d\mu. \tag{5}$$

Apply the operator $\widetilde{\Delta}$ to both sides of equality (5). Using the fact that the operators S_w commute with $\widetilde{\Delta}$, the Euler-Poisson-Darboux equation $\widetilde{\Delta}_w (S_w f)(z) = \widetilde{\Delta}_z (S_w f)(z)$, and also the fact that $\widetilde{\Delta} f \in C_0(B)$ (and hence $S_w \widetilde{\Delta} f \in C_0(B)$), one obtains

$$\widetilde{\Delta} a_{\mu, f}(w) = \int_B \widetilde{\Delta}_w (S_w f)\, d\mu = \int_B S_w(\widetilde{\Delta} f)\, d\mu = \int_{\overline{B}} S_w(\widetilde{\Delta} f)\, d\mu = 0.$$

The last integral is equal to zero, because $S_w \widetilde{\Delta} f \in Z \subset Y$ and $\mu \in Y^{\perp}$.

Therefore, the function $a_{\mu, f}$ is $\mathscr{M}$-harmonic. But $a_{\mu, f}$ is radial, so, by the maximum principle, $a_{\mu, f} = \text{const}$. Since $a_{\mu, f}(0) = \int_{\overline{B}} f\, d\mu = 0$, one concludes that $a_{\mu, f}(w) = 0$, $w \in B$.

Thus, $S_w f$ is annihilated by all measures $\mu \in Y^{\perp}$, and consequently $S_w f \in Y$. Since the function $f \in Y^{\omega} \cap W$ is arbitrary, one has that $S_w : Y^{\omega} \cap W \to Y$. Lemma 1 implies that $S_w : Y \to Y$. The lemma is proved.

Now let us turn to the proof of the theorem itself. Let $f \in Y$. By Lemma 2, the spherical means

$$(S_w f)(z) = \int_{U(n)} f(\varphi_z(uw))\, du, \qquad |w| < 1,$$

belong to the space Y. Set $w = re_1$, and let $r \to 1$. We obtain

$$\begin{aligned}\lim_{r\to 1}(S_{re_1} f)(z) &= \int_{U(n)} f(\varphi_z(ue_1))\, du = \int_S f(\varphi_z(s))\, d\sigma_S(s) \\ &= (P \circ j)(f \circ \varphi_z)(0) = [(P \circ j)f](z), \qquad z \in B.\end{aligned}$$

In the last equality we used the fact that the Poisson operator P commutes with the automorphism in $\mathcal{M}$.

Therefore, the family $S_{re_1} f$ converges pointwise to the function $(P \circ j)f$ when $r \to 1$. Since the family $S_{re_1} f$ is uniformly bounded and the limit function belongs to $C(\overline{B})$, Lebesgue's dominated convergence theorem implies that $(P \circ j)f$ is orthogonal to any measure in the annihilator of the space Y and hence $(P \circ j)f \in Y$. The theorem is proved.

REMARK 1. The theorem remains valid if instead of closed $\mathcal{M}$-invariant subspaces $Y \subset C(\overline{B})$, one considers closed $\mathcal{M}$-invariant subspaces of the space W, equipped with the norm $\|f\|_W = \|f\|_{C(\overline{B})} + \|\widetilde{\Delta} f\|_{L^2(B,\tau)}$.

REMARK 2. Let $n = 1$. One can show that any closed $\mathcal{M}$-invariant subspace $Y \subset C(\overline{B})$ possesses a dense subset consisting of twice differentiable (in the open unit disk) functions, for which the ordinary Laplacian satisfies the inequality

$$|\Delta f(z)| \le C(1 - |z|)^{-2}.$$

The assumptions of the theorem proved above require the existence of a dense subset of functions $f \in C^2(B)$ for which $\Delta f \in L^2(B, (1-|z|^2)^2 d\sigma_B)$, for instance, functions whose Laplacian is subject to the estimate

$$|\Delta f(z)| \le C(1 - |z|)^{-3/2+\varepsilon}, \qquad \epsilon > 0.$$

Therefore, the theorem establishes the validity of the decomposition (1) under restriction on the rate of growth of the Laplacian as one approaches the boundary of the disk that are somewhat stronger than those automatically satisfied in arbitrary Möbius-invariant spaces.

§7. Möbius $A(B^1)$-modules in the unit disk

In this section we present a complete description of all the Möbius $A(B^1)$-modules in the closed unit disk $\overline{B}^1$, i.e., closed Möbius subspaces $Y \subset C(\overline{B}^1)$ such that $hf \in Y$ for any $f \in Y$ and every $h \in A(B^1)$ (this is clearly equivalent to the condition that $zf(z) \in Y$ for any $f \in Y$).

7.1. THEOREM. *Every closed Möbius $A(B^1)$-submodule of $C(\overline{B}^1)$ is one of the entries in the following diagram*

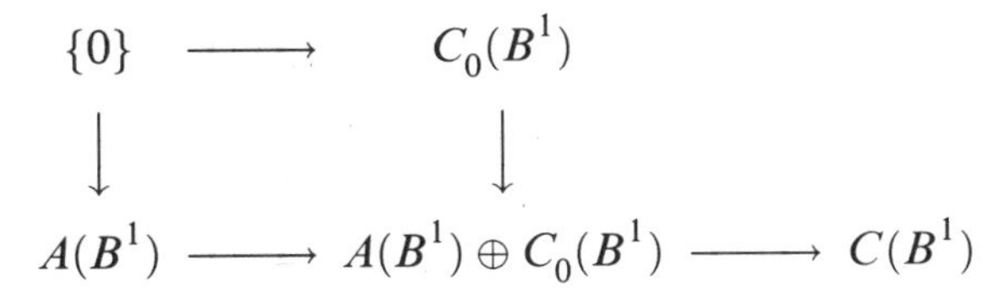

$$\begin{array}{ccccc} \{0\} & \longrightarrow & C_0(B^1) & & \\ \downarrow & & \downarrow & & \\ A(B^1) & \longrightarrow & A(B^1)\oplus C_0(B^1) & \longrightarrow & C(B^1) \end{array}$$

We first prove a few lemmas. Throughout this section, Y denotes a closed Möbius $A(B^1)$-submodule of $C(\overline{B}^1)$, $Y \neq \{0\}$. Set $Y_0 = Y \cap C_0(B^1)$.

7.2. LEMMA. *Assume that $Y_0 \neq Y$. Then $A(B^1) \subset Y$.*

PROOF. It suffices to prove that Y contains all the constant functions. For $\mu \in Y^\perp$, let $\widetilde{\mu}$ denote the mean value of μ over the rotation group:

$$\widetilde{\mu} = \frac{1}{2\pi}\int_0^{2\pi} (\mu \circ k_\theta)\, d\theta,$$

where k_θ is rotation on the angle θ.

By assumption, there exists $f \in Y$ which does not vanish from zero at some point $z_0 \in \partial B^1$. For any $t \in [0, 1)$,

$$\int_{\overline{B}^1} f\left(\frac{z+tz_0}{1+t\overline{z}_0 z}\right) d\widetilde{\mu}(z) = 0.$$

Now, for any $z \in \overline{B}^1$, $z \neq -z_0$,

$$\lim_{t\to 1} \frac{z+tz_0}{1+t\overline{z}_0 z} = z_0,$$

and $\{-z_0\}$ is the null set $\widetilde{\mu}$; hence, $f(z_0)\int_{\overline{B}^1} d\widetilde{\mu} = 0$. Therefore, $\int_{\overline{B}^1} d\mu = \int_{\overline{B}^1} d\widetilde{\mu} = 0$. Since $\mu \in Y^\perp$ is arbitrary, this means that $\mathbb{C} \subset Y$.

7.3. LEMMA. *Assume that $Y_0 = C_0(B^1)$. Then Y is one of the spaces $C_0(B^1)$, $A(B^1) \oplus C_0(B^1)$, $C(\overline{B}^1)$.*

PROOF. Let $\mu \in Y^\perp$. Since $C_0(B^1) \subset Y$, the support of μ is contained in the circle ∂B^1. The assertion now follows from the fact that the uniform closure $\mathrm{cl}\left[Y|_{\partial B^1}\right]$ of the restriction of Y to ∂B^1 is one of the spaces $\{0\}$, $A(\partial B^1)$, $C(\partial B^1)$.

7.4. In this section we will construct a differential operator that maps Y into the space of functions integrable with respect to an invariant measure τ.

The group $\mathscr{M}$ of conformal transformations of the unit disk is isomorphic to the group of complex matrices

$$\omega = \begin{pmatrix} \alpha & \beta \\ \overline{\beta} & \overline{\alpha} \end{pmatrix}, \qquad |\alpha|^2 - |\beta|^2 = 1,$$

whose action in the unit disc is described by the formula

$$\omega z = \frac{\alpha z + \beta}{\overline{\beta} z + \overline{\alpha}}.$$

Consider the following family of operators:

$$(R_\omega f)(z) = \frac{1}{(-\overline{\beta} z + \alpha)^2} f(\omega^{-1} z).$$

These operators define a unitary representation R of $\mathscr{M}$ in $L^2(B^1, d\overline{z}dz)$. Consider also the differential operator

$$\overline{\partial} = (1 - |z|^2)^2 \frac{\partial}{\partial \overline{z}}.$$

This is an intertwining operator of R and the quasi-regular representation T of $\mathscr{M}$ in $L^2(B^1, \tau)$, as defined by translation operators:

$$\overline{\partial} R_\omega = T_\omega \overline{\partial}, \qquad \omega \in \mathscr{M}. \tag{1}$$

It is clear that the operators R_ω map Y into itself, since Y is a Möbius space and an $A(B^1)$-module. Now we want to show that $\overline{\partial}$ maps the space $Y^\omega \subset Y$ of analytic vectors into the space $L^2(B^1, \tau)$.

Let $f \in Y^\omega$. Denote by $\omega_{\theta,2}$ the following element of $\mathscr{M}$:

$$\omega_{\theta,\alpha}(z) = e^{i\theta} \frac{z + \alpha}{1 + \overline{\alpha} Z}, \qquad \theta \in [0, 2\pi), \ |\alpha| < 1.$$

Set $f_{\theta,\alpha} = f \circ \omega_{\theta,\alpha}$.

Since f is an analytic vector of the representation T, it follows that $f_{\theta,\alpha}$ can be expanded in the convergent power series in parameters $\theta, \alpha, \overline{\alpha}$ of the group $\mathscr{M}$:

$$f_{\theta,\alpha} = \sum_{k,m,n \geq 0} f_{k,m,n} \frac{\theta^k}{k!} \frac{\alpha^m}{m!} \frac{\overline{\alpha}^n}{n!},$$

with coefficients $f_{k,m,n}$ in $Y \subset C(\overline{B}^1)$. In particular,

$$\left.\frac{\partial}{\partial \alpha}\right|_{\substack{\theta=0\\ \alpha=0}} f_{\theta,\alpha} = f_{0,1,0} \in C(\overline{B}^1),$$

$$\left.\frac{\partial}{\partial \overline{\alpha}}\right|_{\substack{\theta=0\\ \alpha=0}} f_{\theta,\alpha} = f_{0,1,0} \in C(\overline{B}^1).$$

On the other hand, one has

$$\left.\frac{\partial}{\partial \alpha}\right|_{\substack{\theta=0\\ \alpha=0}} f_{\theta,\alpha}(z) = \left.\frac{\partial}{\partial \alpha}\right|_{\alpha=0} f\left(\frac{z + \alpha}{1 + \overline{\alpha} z}\right) = \frac{\partial f}{\partial z} - \overline{z}^2 \frac{\partial f}{\partial \overline{z}},$$

$$\left.\frac{\partial}{\partial \overline{\alpha}}\right|_{\substack{\theta=0\\ \alpha=0}} f_{\theta,\alpha}(z) = \left.\frac{\partial}{\partial \overline{\alpha}}\right|_{\alpha=0} f\left(\frac{z + \alpha}{1 + \overline{\alpha} z}\right) = \frac{\partial f}{\partial \overline{z}} - z^2 \frac{\partial f}{\partial z}.$$

The last four identities imply

$$\frac{\partial f}{\partial \overline{z}} = \frac{z^2 f_{0,1,0} + f_{0,0,1}}{1 - |z|^4},$$

and therefore

$$(1 - |z|^2)\frac{\partial f}{\partial \overline{z}} = \frac{z^2 f_{0,1,0} + f_{0,0,1}}{1 + |z|^2}$$

is a bounded function in B^1. Since $d\tau = (1 - |z|^2)^{-2} d\sigma$, we obtain

$$\int_{B^1} |\overline{\partial} f|^2 \, d\tau = \int_{B^1} (1 - |z|^2)^2 \left|\frac{\partial f}{\partial \overline{z}}\right|^2 d\sigma < \infty,$$

i.e., $\overline{\partial} f \in L^2(B^1, \tau)$. Thus $\overline{\partial}(Y^\omega) \subset L^2(B^1, \tau)$.

7.5. LEMMA. *If $Y_0 \neq \{0\}$, then $Y_0 = C_0(B^1)$.*

PROOF. Consider the space Y^ω of analytic vectors. By Lemma 6.1.2, $\widetilde{\Delta} f \in Y_0$ for any $f \in Y^\omega$. Moreover, $zf \in Y^\omega$. Then

$$(1 - |z|^2)^2 \frac{\partial f}{\partial \overline{z}} = \widetilde{\Delta}(zf) - z\widetilde{\Delta} f \in Y_0.$$

Thus, $\overline{\partial} f \in Y_0$ and also $\overline{\partial} f \in L^2(B^1, \tau)$, as we already know. Hence, $\overline{\partial} Y^\omega \subset Y_0 \cap L^2(B^1, \tau)$. Set $E = \mathrm{cl}_{L^2(B^1, \tau)} \left[\overline{\partial} Y^\omega\right]$. Since Y^ω is invariant under the operators R_ω, it follows from the commutation relation (1), §7.4, that E is invariant under the operators T_ω; in other words, E is a Möbius subspace.

By Theorem 3.7, Chapter I, E is closed under complex conjugation. Moreover, $zE \subset E$, because $zY^\omega \subset Y^\omega$ and $\overline{\partial}$ commutes with multiplication by z. Therefore, $\overline{z}E = \overline{z\overline{E}} = \overline{zE} \subset \overline{E} = E$; i.e., E is invariant under multiplication by z and $\overline{z}$, hence also by an arbitrary function in $C(\overline{B}^1)$. Therefore E is the space of all elements of $L^2(B^1, \tau)$ that vanish a.e. on a fixed measurable set $\eta \subset B^1$. If $\operatorname{mes} \eta > 0$, then by invariance $E = \{0\}$. But then $\overline{\partial} Y^\omega = \{0\}$, and all the elements of Y^ω, hence also of Y, are holomorphic functions. But this contradicts the maximum principle and the assumption $Y_0 \neq \{0\}$. Thus, $\operatorname{mes} \eta = 0$, i.e., $E = L^2(B^1, \tau)$.

Lemma 3.1, Chapter I, implies that

$$L^2(B^1, \tau) \cap C_0(B^1) \subset \mathrm{cl}_{C(\overline{B}^1)} \left[\overline{\partial} Y^\omega\right] \subset Y_0,$$

which in turn implies that $Y_0 = C_0(B^1)$.

7.6. PROOF OF THEOREM 7.1. Suppose that $Y_0 \neq \{0\}$. By Lemma 7.5, $Y_0 = C_0(B^1)$, and, by Lemma 7.3, Y is one of the spaces $C_0(B^1)$, $A(B^1)\oplus$

$C_0(B^1)$, $C(\overline{B}^1)$. If $Y_0 = \{0\}$, Theorem 5.1 implies that all functions in Y are harmonic. Since $zY \subset Y$, all of them are holomorphic, i.e., $Y \subset A(B^1)$. On the other hand, since $Y \neq Y_0 = \{0\}$, we have $A(B^1) \subset Y$ by Lemma 7.2. So, $Y_0 = \{0\}$ implies $Y = A(B^1)$, and this completes the proof.

We now illustrate the application of Theorem 7.1.

7.7. PROPOSITION. *Let μ be a finite nonzero Borel measure in the open unit disk B^1. Let $f \in C(B^1)$ be such that*

$$\int_{|z|<1} z^k f(\omega(z))\, d\mu = 0, \qquad k = 0, 1, \dots,$$

for all $\omega \in \mathcal{M}$. Then $f \in A(B^1)$.

PROOF. The space Y of all functions f satisfying the assumption is a closed $A(B^1)$-module contained in $C(\overline{B}^1)$. Since $\mu \neq 0$, it follows that $C_0(B^1) \not\subset Y$. Therefore, either $Y = \{0\}$ or $Y = A(B^1)$.

7.8. EXAMPLE. Let μ be the measure dz with support on a piecewise smooth contour $\gamma \subset B^1$. By the Golubev-Privalov theorem, the assumption on f implies the existence of an analytic continuation of f from any contour $\omega(\gamma)$, $\omega \in \mathcal{M}$, into the domain bounded by the contour. We do not know in advance whether this continuation coincides with f inside $\omega(\gamma)$. Proposition 7.7 states that in this case f is a holomorphic function; therefore, the holomorphic extension inside $\omega(\gamma)$ is the function f itself.

The following similar holomorphy test in a disk uses conformal transformations.

7.9. PROPOSITION. *Let F be a compact subset of B^1. Let $P(F)$ denote the closure of the space of polynomials in z in the topology of uniform convergence in F. Assume that $P(F) \neq C(F)$. Let $f \in C(\overline{B}^1)$ be such that $f \in P(\omega(F))$ for all $\omega \in \mathcal{M}$. Then $f \in A(B^1)$.*

The proof is the same as in §7.7.

7.10. Remarks to Chapter III. Theorem 1.3 for $n = 1$ was proved by de Leeuw and Mirkil [72, §2 of Chapter II]. In the one-dimensional case the condition that an algebra is invariant with respect to dilatations is equivalent to the assumption that it contains an approximate unit.

The results of section 4 were obtained jointly with A. M. Semenov. Nagel and Rudin [84] (see also Rudin's monograph [26]) determined all the closed Möbius subalgebras of $C(\overline{B}^1)$. Previously [2], the author classified the closed Möbius subalgebras of $C(\overline{B}^n)$ that contain nonconstant holomorphic or antiholomorphic functions (in the case $n = 1$ this result was proved in a joint paper with R. E. Val′skiĭ [31]).

Holomorphy criteria in a ball, using the group of analytic automorphisms in the spirit of Propositions 7.7 and 7.9, were proved first in [31] (for disks) and then in [2] (general case),where they appeared as corollaries of the classification of Möbius algebras. In Chapter III such holomorphy criterions are given in the case $n = 1$ as sample applications of the classification of Möbius $A(B^1)$-modules. In the next chapter analogous holomorphy criteria will be proved for multidimensional domains under weaker assumptions.

CHAPTER IV

Holomorphy Tests in Symmetric Domains Involving the Automorphism Group. Related Problems

In this chapter we obtain further results concerning the description of holomorphic functions or their boundary values, using the properties of translations by analytic automorphisms.

§1. Integral conditions for the existence of a holomorphic extension from the boundary of a symmetric domain

In the present section we will use the descriptions of Möbius spaces on the Shilov boundary of a symmetric domain, obtained in Chapter III, §3, to characterize functions of $A(\partial D)$ in terms of integrals over ∂D.

1.1. PROPOSITION. *Let D be a bounded irreducible symmetric domain considered as a disk domain in $\mathbb{C}^n$. Let μ be a Borel measure on ∂D such that*

$$\int_{\partial D} \overline{z}_k \, d\mu \neq 0$$

for some k. If $f \in C(\partial D)$ is such that

$$\int_{\partial D} (f \circ \omega) \, d\mu = 0$$

for all $\omega \in \operatorname{Aut}(D)$, then $f \in A(\partial D)$.

PROOF. The set Y of functions f satisfying the assumption is a closed Möbius subspace of $C(\partial D)$, and $\overline{z}_k \notin Y$. By Theorem 3.1, $Y \subset A(\partial D)$.

Picking a particular μ, we obtain the following

1.2. PROPOSITION. *Pick some $z_0 \in \partial D$. Let $f \in C(\partial D)$ be such that*

$$\int_0^{2\pi} f\left(\omega(e^{i\theta} z_0)\right) e^{i\theta} \, d\theta = 0$$

for all $\omega \in \operatorname{Aut}(D)$. Then $f \in A(\partial D)$.

1.3. PROPOSITION. *Let* $f \in C(\partial D)$. *Assume that for some* $k = 1, \dots, n$ *and all* $\omega \in \operatorname{Aut}(D)$

$$\int_{\partial D} f(z)[\omega(z)]_k P(\omega^{-1}(0), z)\, d\sigma(z) = 0,$$

where $P(\zeta, z)$ *is the invariant Poisson kernel of the symmetric domain* D, σ *a measure invariant under the stationary subgroup of zero in* $\operatorname{Aut}(D)$, *and* $[\omega(z)]_k$ *the* k *th coordinate of the map* $\omega : D \to D$. *Then* $f \in A(\partial D)$.

PROOF. Change the integration variable $z = \omega^{-1}(w)$. Since the Jacobian of this change is the Poisson kernel, we have

$$\int_{\partial D} f(\omega^{-1}(w)) w_k \, d\sigma(w) = 0, \qquad \omega \in \operatorname{Aut}(D).$$

The measure $\mu = w_k \sigma$ satisfies the assumptions of Proposition 1.1; therefore, $f \in A(\partial D)$.

As an illustration of Proposition 1.3, let us consider, as an example, the generalized disk $D^I_{n,n}$. The Poisson kernel is then given by the formula [30]

$$P(\zeta, z) = c \frac{[\det(e - \zeta^* \zeta)]^n}{|\det(\zeta - w)|^{2n}},$$

and σ is the Haar measure on the unitary group $U(n) = \partial D^I_{n,n}$. Any automorphism ω of the domain $D^I_{n,n}$ can be represented as

$$\omega(z) = a(z - \lambda)(e - \lambda z)^{-1} b,$$

where $\lambda = \operatorname{diag}\{\lambda_1, \dots, \lambda_n\}$, $|\lambda_i| < 1$, a and b are constant unitary matrices. This argument, together with Proposition 1.3, proves the following result.

1.4. PROPOSITION. *Let* $f \in C(\partial D^I_{n,n})$. *Then* $f \in A(\partial D^I_{n,n})$ *if and only if the following vanishes:*

$$\int_{U(n)} f(w) \frac{(w - \lambda)(e - \lambda w)^{-1}}{|\det(\lambda - w)|^{2n}} \, d\sigma(w) = 0,$$

where λ *is an arbitrary diagonal matrix such that* $|\lambda_i| < 1$.

The function

$$F(z, \lambda) = \frac{(z - \lambda)(e - \lambda z)^{-1}}{|\det(\lambda - z)|^{2n}}$$

is the generating function of the family of polynomials

$$P_I(z) = \left. \frac{\partial^{|I|}}{\partial \lambda^I} F(z, \lambda) \right|_{\lambda = 0} = I! \sum_{K+L+M=I} c_{K,L,M} q_k(z) \Delta^L(z) \overline{\Delta^M(z)},$$

where $I = (i_1, \dots, i_n)$ and

$$c_{K,L,M} = (-1)^{|L|+|M|} \frac{(k+|L|-1)!(n+|M|-1)!}{[(n-1)!]^2 L! M!}, \qquad \Delta^L(z) = \Delta_1^{l_1}(z) \cdots \Delta_n^{l_n}(z).$$

$\Delta_i(z)$ is the cofactor of the element z_{ii} of the matrix z, and $q_K(z)$ is a polynomial matrix which is the coefficient of λ^K in the expansion of the matrix $(\lambda z)^{|K|} - \lambda(\lambda z)^{|K|-1}$ in powers of λ_i. This proves the following proposition.

1.5. Proposition. *A function $f \in C(U(n))$ admits a continuous holomorphic extension into the domain $D^I_{n,n} = \{e - z^* z > 0\}$ if and only if*

$$\int_{\partial D} f(z) P_I(z)\, d\sigma(z) = 0$$

for any polynomial P_I.

Analogous conditions for the existence of analytic extensions can be derived in explicit form for other classical domains.

Remark. L. A. Aĭzenberg and Sh. A. Dautov [32] obtained other polynomial families with the same property. Propositions 1.4–1.6 are also valid for functions $f \in L^p(\partial D, \sigma)$. In that case they provide necessary and sufficient conditions for a function to belong to the space $H^p(\partial D) = H^p(\partial D, \sigma)$.

1.6. Proposition. *Let $f \in L^p(\partial D)$ be such that the Cauchy-Bergmann integral (see, e.g., [30]) of f commutes with the analytic automorphisms of D, i.e.,*

$$C[f \circ \omega] = C[f] \circ \omega, \qquad \omega \in \operatorname{Aut}(D).$$

Then $f \in H^p(\partial D)$.

Proof. Consider D as a disk domain. The functions f satisfying the assumption form a closed Möbius subspace Y of $L^p(\partial D, \sigma)$ such that $H^p(\partial D) \subset Y$. If we prove that Y does not contain any nonconstant functions of $\overline{H}^p(\partial D)$, it will follow by Theorem 3.1, Chapter III, that $Y \subset H^p(\partial D)$.

Let $g \in Y \cap \overline{H}^p(\partial D)$. Since the Cauchy-Bergmann kernel is antiholomorphic with respect to its second argument, we have

$$C[g](w(z)) = \int_{\partial D} C(w(z), \zeta) g(\zeta)\, d\sigma(\zeta) = C(w(z), 0) g(0) = g(0),$$

$$C[g \circ \omega](z) = \int_{\partial D} C(z, \zeta) g(\omega(\zeta))\, d\sigma(\zeta) = g(\omega(0)),$$

and the transitivity of the group $\operatorname{Aut}(D)$ implies $g = \text{const}$.

§2. Morera theorems in the unit disk for conformally invariant families of contours

In §6.8, Chapter III, we showed that a function f is holomorphic in the open unit disk B^1 if it has a holomorphic extension inside every closed contour in a certain conformally invariant family. This followed from the description of closed Möbius $A(B^1)$-modules in $C(\overline{B}^1)$ (Theorem 6.1).

In the present section we will use the same ideas as in the proof of Theorem 6.1 to show that, instead of the existence of holomorphic extensions, it will suffice to assume that the integrals over all contours in a certain conformally invariant family vanish. The "size" of this family will depend on the function class in question.

2.1. THEOREM. *Let $f \in L^2(B^1, d\overline{z}dz)$. Suppose that for some closed piecewise smooth Jordan curve $\gamma \subset B^1$ and almost all (with respect to Haar measure) $\omega \in \operatorname{Aut}(B^1)$*

$$\int_{\omega(\gamma)} f(z)\,dz = 0.$$

Then f coincides with a holomorphic function almost everywhere.

We need some preparations for the proof of this theorem.

2.2. Consider $\mathscr{M} = \operatorname{Aut}(B^1)$ as the group of matrices

$$\omega = \begin{pmatrix} \alpha & \beta \\ \overline{\beta} & \overline{\alpha} \end{pmatrix}, \qquad |\alpha|^2 - |\beta|^2 = 1,$$

acting in the disk by the formula

$$\omega z = \frac{\alpha z + \beta}{\overline{\beta} z + \overline{\alpha}}.$$

We consider $\mathscr{M}$ in the standard parametrization:

$$\omega_{\theta, s} = e^{i\theta}\frac{z+s}{1+\overline{s}z}, \qquad \theta \in [0, 2\pi), \ |s| < 1.$$

If $\omega = \omega_{\theta, s}$, we write $\theta = \theta(\omega)$ and $s = s(\omega)$. We write $\omega_z = \omega_{0, z}$. The element ω_z is represented by the matrix

$$\omega_z = \begin{pmatrix} (1-|z|^2)^{-1/2} & z(1-|z|^2)^{-1/2} \\ \overline{z}(1-|z|^2)^{-1/2} & (1-|z|^2)^{-1/2} \end{pmatrix}.$$

Let L^2_{hol} denote the subspace consisting of holomorphic functions in

$$L^2(B^1) = L^2(B^1, d\overline{z}\,dz),$$

and τ the $\mathscr{M}$-invariant measure $d\tau = (1-|z|^2)^{-1/2} d\overline{z}\,dz$.

2.3. Let γ be a closed piecewise smooth Jordan curve contained in B^1, and $H(\gamma)$ the subspace of $L^2(B^1)$ consisting of all f such that

$$\int_{\omega(\gamma)} f(z)\,dz = 0 \tag{1}$$

for almost all $\omega \in \mathscr{M}$. By the Cauchy theorem, $L^2_{\text{hol}} \subset H(\gamma)$. Let

$$H(\gamma) = L^2_{\text{hol}} \oplus E(\gamma)$$

be an orthogonal decomposition.

Let $f \in H(\gamma)$. Changing the variable $w = \omega z = (\alpha z + \beta)/(\overline{\beta} z + \overline{\alpha})$ in the integral (1), we see that $H(\gamma)$ contains the function

$$(R_\omega f)(z) = r(\omega, z) f(\omega^{-1} z), \qquad r(\omega, z) = \frac{1}{(-\overline{\beta} z + \alpha)^2}.$$

It is easily proved that the operators R_ω, $\omega \in \mathscr{M}$, define a unitary representation of $\mathscr{M}$ in $L^2(B^1, \tau)$. Since $H(\gamma)$ and L^2_{hol} are invariant spaces for this representation, it follows that $E(\gamma)$ is an invariant space for the operators R_ω as well.

Consider the operator $\overline{\partial}$ introduced in §7.4, Chapter III:

$$\overline{\partial} = (1 - |z|^2)^2 \frac{\partial}{\partial \overline{z}}.$$

We have already noted that it is an intertwining operator for the representations R_ω and T_ω, i.e., $\overline{\partial} R_\omega = T_\omega \overline{\partial}$.

2.4. Lemma. *The domain $E_{\overline{\partial}}$ of $\overline{\partial}$ is dense in $E(\gamma)$. The range of $\overline{\partial}$ is contained in $L^2(B^1, \tau)$, and it is invariant with respect to the translation operators T_ω.*

Proof. For any compactly supported C^∞-smooth function φ on $\mathscr{M}$, set

$$R(\varphi) f = \int_{\mathscr{M}} \varphi(\omega)\,(R_\omega f)\,d\omega, \qquad f \in E(\gamma). \tag{1}$$

Clearly $R(\varphi) f \in E(\gamma)$. Let $E_{\overline{\partial}}$ denote the linear span of the functions $R(\varphi) f$, $\varphi \in C_c^\infty(\mathscr{M})$, $f \in E(\gamma)$. Then $E_{\overline{\partial}}$ is dense in $E(\gamma)$, and it follows from the equality

$$R(\varphi) f = \int_{\mathscr{M}} \varphi(\omega_z \eta^{-1}) f(\eta(0)) r(\omega_z \eta^{-1}, z)\,d\eta$$

that $R(\varphi) f \in C^\infty(B^1)$. It remains to prove that $\overline{\partial} R(\varphi) f \in L^2(B^1, \tau)$. We have

$$\frac{\partial}{\partial \overline{z}} R(\varphi) f(z) = \int_{\mathscr{M}} \left[\frac{\partial}{\partial \overline{z}} \varphi\left(\omega_z \eta^{-1}\right) r(\omega_z \eta^{-1}) \right.$$

$$\left. + \varphi\left(\omega_z \eta^{-1}\right) \frac{\partial}{\partial \overline{z}} r(\omega_z \eta^{-1}) \right] f(\eta(0))\,d\eta. \tag{2}$$

We estimate each term separately. Fix some $\eta \in \mathscr{M}$, and set $\theta_0 = \theta(\eta^{-1})$, $s_0 = s(\eta^{-1})$. Then

$$e^{i\theta(\omega_z\eta^{-1})} = \frac{e^{i\theta_0} + \overline{s}_0 z}{1 + e^{i\theta_0} s_0 \overline{z}},$$

$$s(\omega_z\eta^{-1}) = \frac{e^{i\theta_0} s_0 + z}{e^{i\theta_0} + \overline{s}_0 z}.$$

Therefore,

$$\frac{\partial}{\partial \overline{z}} \theta(\omega_z\eta^{-1}) = \frac{i e^{i\theta_0} s_0}{1 + e^{i\theta_0} s_0 \overline{z}},$$

$$\frac{\partial}{\partial \overline{z}} \overline{s}(\omega_z\eta^{-1}) = \frac{e^{i\theta_0}(1 - |s_0|^2)}{(1 + e^{i\theta_0} s_0 \overline{z})^2}.$$

Since $|s_0| < 1$, $|z| \le 1$, we have

$$\left|\frac{\partial}{\partial \overline{z}} \theta(\omega_z\eta^{-1})\right| \le \frac{2}{1 - |s_0|^2},$$
$$\left|\frac{\partial}{\partial \overline{z}} \overline{s}(\omega_z\eta^{-1})\right| \le \frac{4}{1 - |s_0|^2}. \tag{3}$$

Since $s_0 = e^{-i\theta_0}\omega_z^{-1}\omega(0)$, it follows that

$$1 - |s_0|^2 = 1 - \frac{|\omega(0) - z|^2}{|1 - \omega(0)\overline{z}|^2} = \frac{(1 - |z|^2)(1 - |\omega(0)|^2)}{|1 - \omega(0)\overline{z}|^2} \ge \frac{1 - |\omega(0)|}{1 + |\omega(0)|}(1 - |z|^2).$$

As the integration domain in (1) is compact, the inequality $|\omega(0)| \le c_1 < 1$ holds in it. The preceding inequality, together with (3), yields

$$\left|\frac{\partial}{\partial \overline{z}} \theta(\omega_z\eta^{-1})\right| \le \frac{2(1 + c_1)}{1 - c_1} \frac{1}{1 - |z|^2},$$

$$\left|\frac{\partial}{\partial \overline{z}} \overline{s}(\omega_z\eta^{-1})\right| \le \frac{4(1 + c_1)}{1 - c_1} \frac{1}{1 - |z|^2}.$$

Hence

$$\begin{aligned}
\left|\frac{\partial}{\partial \overline{z}} \varphi(\omega_z\eta^{-1})\right| &= \left|\frac{\partial \widetilde{\varphi}}{\partial \theta}\left(\theta(\omega_z\eta^{-1}), s(\omega_z\eta^{-1})\right) \cdot \frac{\partial}{\partial \overline{z}} \theta(\omega_z\eta^{-1})\right. \\
&\quad \left. + \frac{\partial \widetilde{\varphi}}{\partial s}\left(\theta(\omega_z\eta^{-1}), s(\omega_z\eta^{-1})\right) \cdot \frac{\partial}{\partial \overline{z}} \overline{s}(\omega_z\eta^{-1})\right| \\
&\le \frac{2(1 + c)}{1 - c} \frac{1}{1 - |z|^2} \left\{\left|\frac{\partial \widetilde{\varphi}}{\partial \theta}\left(\theta(\omega_z\eta^{-1}), s(\omega_z\eta^{-1})\right)\right|\right. \\
&\quad \left. + 2\left|\frac{\partial \widetilde{\varphi}}{\partial \overline{s}}\left(\theta(\omega_z\eta^{-1}), s(\omega_z\eta^{-1})\right)\right|\right\},
\end{aligned} \tag{4}$$

where $\widetilde{\varphi}(\theta, s) = \varphi(\omega_{\theta,s})$.

Now let us estimate the derivative of $r(\omega_z\eta^{-1}, z)$. Since

$$r(\omega_z\eta^{-1}, z) = (1-|z|^2)\left[(a\bar{z}-b)z + (a+\bar{b}z)\right]^{-2},$$

where $\eta^{-1} = \begin{pmatrix} a & b \\ \bar{b} & \bar{a} \end{pmatrix}$, it follows that

$$\left|\frac{\partial}{\partial\bar{z}} r(\omega_z\eta^{-1}, z)\right| \le \left|r(\omega_z\eta^{-1}, z)\right| \cdot \frac{1}{1-|z|^2} + 2|a|\frac{\left|r(\omega_z\eta^{-1}, z)\right|^{3/2}}{(1-|z|^2)^{1/2}}.$$

The element a of the matrix η^{-1} can be determined from the equation $\eta^{-1} = \omega_z^{-1}\omega$:

$$a = \frac{\omega_1 - z\overline{\omega}_2}{(1-|z|^2)^{1/2}}, \qquad \omega = \begin{pmatrix} \omega_1 & \omega_2 \\ \overline{\omega}_2 & \overline{\omega}_1 \end{pmatrix}.$$

The elements ω_1, ω_2 and the function $r(\omega_z\eta^{-1}, z)$ are bounded in the domain of integration of (1), because the latter is compact: $|\omega_1|, |\omega_2|, |r| \le c_2$.

Inequality (4) yields

$$\left|\frac{\partial}{\partial\bar{z}} r(\omega_z\eta^{-1}, z)\right| \le c_3(1-|z|^2)^{-1}, \qquad c_3 = c_2 + 4c_2^{5/2}. \tag{5}$$

Substituting (4) and (5) into (2) and going back to the original integration variable, we get

$$\left|\frac{\partial}{\partial\bar{z}} R(\varphi)f(z)\right| \le \frac{1}{1-|z|^2}\cdot\Psi(z), \tag{6}$$

where $\Psi(z) = \int_{\mathcal{M}} \psi(\omega)|f(\omega^{-1}z)|\,d\omega$ and ψ is a function compactly supported and continuous in $\mathcal{M}$. Since $f \in L^2(B^1, d\bar{z}dz)$, we have $\Psi \in L^2(B^1, d\bar{z}dz)$. By (6), this implies $\bar{\partial}R(\varphi)f \in L^2(B^1, \tau)$. The invariance of the range of $\bar{\partial}$ follows from the relation $\bar{\partial}R_\omega = T_\omega\bar{\partial}$.

2.5. Proof of Theorem 2.1. Let $E_1(\gamma)$ denote the subspace of $E(\gamma)$ consisting of the functions f such that

$$f(e^{i\theta}z) = e^{-i\theta}f(z), \qquad \theta \in [0, 2\pi).$$

Consider the projection operator

$$(S_1 f)(z) = \frac{1}{2\pi}\int_0^{2\pi} f\left(e^{i\theta}z\right)e^{i\theta}\,d\theta$$

of $E(\gamma)$ onto the subspace $E_1(\gamma)$. It follows from Lemma 2.4 that $E_{\bar{\partial}} \cap E_1(\gamma)$ is dense in $E_1(\gamma)$.

Let $f \in E_{\overline{\partial}} \cap E_1(\gamma)$. Applying Green's formula to the equality

$$\int_{\omega(\gamma)} f(z)\,dz = 0,$$

we obtain

$$\int_{\omega(\Omega)} \frac{\partial f}{\partial \overline{z}}\,d\overline{z} \wedge dz = 0,$$

where Ω is the domain bounded by γ. Letting χ_Ω denote the characteristic function of Ω, we rewrite this equality as

$$\int_{B^1} \overline{\partial} f \cdot \left(\chi_\Omega \circ \omega^{-1}\right) d\tau = 0. \tag{1}$$

Using the canonical projection $\pi : \mathcal{M} \to B^1$, $\pi(\omega) = \omega(0)$, we can lift this equation to $\mathcal{M}$:

$$\int_{\mathcal{M}} (\overline{\partial} f \circ \pi)(\eta)\,(\chi_\Omega \circ \pi)(\omega^{-1}\eta)\,d\eta = 0.$$

We can interpret the integral as a convolution on $\mathcal{M}$:

$$(\overline{\partial} f \circ \pi) * (\widetilde{\chi_\Omega \circ \pi}) = 0,$$

where by definition $\widetilde{h}(\omega) = h(\omega^{-1})$.

The Cayley transform induces an isomorphism of $\mathcal{M}$ and the unimodular group $\mathrm{SL}_2(\mathbb{R})$. Thus, we can represent the last equality in terms of convolutions on $\mathrm{SL}_2(\mathbb{R})$:

$$f' * \chi' = 0. \tag{2}$$

The function χ' is compactly supported, because χ_Ω has this property.

Let K denote the group $\mathrm{SO}(2)$. Since $(\overline{\partial} f)(e^{i\theta} z) = \overline{\partial} f(z)$, it follows that f' is bi-invariant with respect to K. In addition, Lemma 2.4 implies that $f' \in L^2(\mathrm{SL}_2(\mathbb{R}))$.

Taking the K-spherical Fourier transform of (2) and noting that one of the factors is K-invariant, we get

$$\widehat{f}' \cdot \widehat{\chi}' = 0.$$

By the Harish-Chandra formula ([1]) already used in §1, Chapter I,

$$\widehat{g}(\lambda) = \widehat{Vg}^K(\lambda), \qquad g \in L^2(\mathrm{SL}_2(\mathbb{R})),\ \lambda \in \mathbb{R},$$

whether the right-hand side is the usual Fourier transform of the function

$$(Vg^K)(t) = e^{-t} \int_{-\infty}^{\infty} \widetilde{g}^K(x, e^t)\,dx,$$

([1]) See S. Lang, $\mathrm{SL}_2(\mathbb{R})$, Addison-Wesley, Reading, MA, 1975.

and $\widetilde{g}^K$ is the pullback to the upper half-plane $\Pi^+ = \mathrm{SL}_2(\mathbb{R})/K$ of the mean value

$$g^K(\omega) = \int_{K\times K} g(k_1\omega k_2)\,dk_1\,dk_2.$$

Since $V\chi'^K$ has compact support, $\widehat{\chi}'$ can be extended analytically to an entire function. Therefore, the measure of the set of zeros of $\widehat{\chi}'$ is zero. By (3), $\widehat{f}' = 0$ almost everywhere. The Plancherel measure for the Fourier transform on $\mathrm{SL}_2(\mathbb{R})$ is absolutely continuous with respect to Lebesgue measure in $\mathbb{R}$. Since $f' \in L_2(\mathrm{SL}_2(\mathbb{R}))$ and $\widehat{f'} = 0$ almost everywhere, it follows that $f = 0$.

For the original functions this yields $\overline{\partial} f = 0$. The operator $\overline{\partial}$ is injective on $E(\gamma)$; hence, $f = 0$. Since f was an arbitrary element of $E_{\overline{\partial}} \cap E_1(\gamma)$, it follows that $E_{\overline{\partial}} \cap E_1(\gamma) = \{0\}$.

Let $f \in E_{\overline{\partial}} \cap E(\gamma)$. Using the identity

$$S\overline{\partial} = \overline{\partial} S_1,$$

where $(Sf) = \frac{1}{2}\int_0^{2\pi} f\left(e^{i\theta}z\right)d\theta$, we get $S\overline{\partial} f = 0$. Then $\overline{\partial} f(0) = \left(S\overline{\partial} f\right)(0) = 0$. But this is true for functions $R_\omega f$, $\omega \in \mathcal{M}$, as well. The equality

$$\overline{\partial} R_\omega f = \left(\overline{\partial} f\right) \circ \omega^{-1}$$

(see formula (1), §7.4, Chapter III) now implies $(\overline{\partial} f)(\omega^{-1}(0)) = 0$ for any $\omega \in \mathcal{M}$, i.e., $\overline{\partial} f = 0$. Then $f = 0$, since $\overline{\partial}$ is injective on $E(\gamma)$. Thus, $E_{\overline{\partial}} \cap E(\gamma) = \{0\}$. Since the latter subspace is dense in $E(\gamma)$, $E(\gamma) = \{0\}$. Therefore, $H(\gamma) = L^2_{\mathrm{hol}}$.

2.6. Under certain conditions Theorem 2.1 remains valid even if the intersection $\gamma \cap \partial B^1$ contains exactly one point z_0 (without loss of generality we may assume that $z_0 = 1$). The crucial parameter in this case is the rate at which γ approaches the disk boundary. It can be proved that if z_0 is a cusp-type point of γ:

$$\operatorname{Im} z = (1 - \operatorname{Re} z)^\alpha, \quad \operatorname{Im} z > 0;$$
$$\operatorname{Im} z = (1 - \operatorname{Re} z)^\alpha, \quad \operatorname{Im} z < 0,$$

where $\alpha > 1$, then the Fourier transform $\widehat{\chi}_\Omega(\lambda)$, $\lambda \in \mathbb{R}$, can be extended analytically into the strip $\{\lambda + i\mu : -\alpha < \mu < \alpha - 1\}$, and the proof holds true.

The vanishing condition of Theorem 2.1 for the integrals over contours belonging to one $\mathcal{M}$-invariant family does not suffice for the analyticity of f, if the space to which f belongs a priori is enlarged. Consider the case when the contours are circles. The following theorem is analogous to J. Delsart's "two radius theorem" for harmonic functions [61], [62], and also to Zalcman's analyticity criterion [106].

2.7. Theorem. *Let $f \in L^1_{\mathrm{loc}}(B^1, d\bar{z}\,dz)$. Suppose there exist two circles $C_{r_i} = \{|z| = r_i\}$, $i = 1, 2$, $0 < r_1, r_2 < 1$, such that*

$$\int_{\omega(C_{r_1})} f(z)\,dz = \int_{\omega(C_{r_2})} f(z)\,dz \tag{1}$$

for almost all $\omega \in \mathcal{M}$. Set $J_r(\lambda) = F(2 - i\lambda, \frac{3}{2}, 3, -4r(1-r)^{-2})$, where F is the hypergeometric function. Then, as soon as J_{r_1}, J_{r_2} have no common zeros in the complex plane, f coincides almost everywhere in B^1 with a holomorphic function.

Proof. First let $f \in C^\infty(B^1)$. As in the proof of Theorem 2.1, we can rewrite condition (1) as

$$\int_{B^1} \bar{\partial} f \cdot \left(\chi_{r_i} \circ \omega^{-1}\right) d\tau = 0, \qquad i = 1, 2, \tag{2}$$

where χ_{r_i} is the characteristic function of the disk $|z| < r_i$. The Harish-Chandra formula for the Fourier transform yields

$$\chi_r(\lambda) = \frac{2\pi r^2 (1-r)^{i\lambda - 4}}{(1+r)^{i\lambda}} F(2 - i\lambda, 3/2, 3, -4r(1-r)^{-2}).$$

It follows from (1) that $\widehat{\chi}_{r_1}$ and $\widehat{\chi}_{r_2}$ have no common zeros in the complex plane.

Let $\mathscr{E}_K$ denote the space of compactly supported K-invariant distributions in B^1 (recall that K is the rotation group). By [99], the set of K-spherical Fourier transforms of the elements of $\mathscr{E}_K$ consists of the entire functions such that

$$\sigma_{n,R}(f) = \sup(1 + |z|)^{-n} e^{-R|\operatorname{Im}|} |f(z)| < \infty.$$

The seminorms $\sigma_{n,R}$ define a topology in $\widehat{\mathscr{E}}_K(B^1)$ in which passage to Fourier transform is a topological isomorphism. This topology coincides with the topology of the space $\widehat{\mathscr{E}}(\mathbb{R})$ of Fourier transforms of compactly supported distributions on the real line.

By a theorem of Schwartz [98] the ideal generated by the functions χ_{r_1}, χ_{r_2} is dense in $\widehat{\mathscr{E}}(\mathbb{R})$. Then the ideal generated by these functions in the space $\widehat{\mathscr{E}}_K(B^1) \subset \widehat{\mathscr{E}}(\mathbb{R})$ of even functions is dense in $\widehat{\mathscr{E}}_K(B^1)$. Then (2) implies $S\bar{\partial} f = 0$. Using translations as in §2.5, we obtain $\bar{\partial} f = 0$, i.e., f is holomorphic.

The case $f \in L^1_{\mathrm{loc}}(B^1, d\bar{z}\,dz)$ can be reduced to the case of smooth functions in the standard manner, using convolutions (in the sense of the representation R_ω) with functions which are smooth and compactly supported on $\mathcal{M}$.

REMARK FOR THE ENGLISH EDITION. Estimate (6), §2.4, shows that if $\varphi \in C_c^\infty(\mathscr{M})$ and if $f \in C^\infty(B^1)$ satisfies the growth estimate

$$|f(z)| \leq C(1-|z|^2)^{-1}, \tag{3}$$

then $\overline{\partial} R(\varphi) f$ belongs to $L^1(B^1, \tau)$. The spherical Fourier transform maps $L^1(B^1, \tau)$ to a space of holomorphic functions in a strip $0 \leq \operatorname{Re}\lambda \leq 1$. Beniamini and Weit [94] have obtained a Tauberian Wiener theorem for this space. They showed that Theorem 2.1 holds true if $f \in L^1_{\text{loc}}$ satisfies estimate (3) and J_{r_1}, J_{r_2} have no common zeros in a strip $-\delta \leq \operatorname{Re}\lambda \leq 1+\delta$, $\delta > 0$.

§3. Holomorphy on invariant families of subsets of symmetric domains

A holomorphy test for symmetric domains in $\mathbb{C}^n$, $n > 1$, analogous to Theorem 2.1, can be proved by reduction to the convolution equation and use of the K-spherical Fourier transform. However, additional assumptions on f are necessary.

3.1. THEOREM. *Let D be a bounded symmetric domain in $\mathbb{C}^n$. Let $f \in C^1(D)$ be such that*

$$\frac{\partial f}{\partial \overline{z}_k} \in L^2\left(D, \frac{\sigma}{\rho}\right),$$

where σ is the Lebesque measure in D, and $\rho(z) = K(z, \overline{z})$, $K(z, \zeta)$ being the Bergmann kernel function of D. Let Ω be a domain with piecewise smooth boundary, such that the closure of Ω is contained in D and for all $\omega \in \operatorname{Aut}(D)$

$$\int_{\omega(\Omega)} f(z)\, dz \wedge d\overline{z}\, [k] = 0, \quad k = 1, \ldots, n. \tag{1}$$

(The differential $d\overline{z}_k$ is omitted.) Then f is holomorphic in D.

PROOF. By the Stokes formula, condition (1) is equivalent to the equality

$$\int_D \frac{1}{\rho} \frac{\partial f}{\partial \overline{z}_k} \cdot \chi_\Omega \circ \omega^{-1} \cdot \rho\, d\sigma = 0,$$

where χ_Ω denotes the characteristic function of Ω.

It follows from our assumptions about the derivatives of f that $(1/\rho)\cdot(\partial f/\partial \overline{z}_k)$ belongs to $L^2(D, \tau)$, where $\tau = \rho\, d\sigma$ is an $\operatorname{Aut}(D)$-invariant measure on D. Thus, to prove that the function is holomorphic, we need only show that the equation

$$\int_D g(z) \chi_\Omega(\omega^{-1} z)\, d\tau(z) = 0 \tag{2}$$

has a unique solution in $L^2(D, \tau)$.

Equation (2) implies the relations

$$\int_D (S_K g)(z)(S_K \chi_\Omega)(\omega^{-1} z)\, d\tau(z) = 0 \tag{3}$$

where S_K is a projection onto the space of K-invariant functions and

$$(S_K g)(z) = \int_K g(kz)\, dk\,,$$

K being the stationary subgroup of some fixed point $o \in D$ in $\mathrm{Aut}(D)$. We can lift this to $\mathrm{Aut}(D)$ as a convolution equation. Take K-spherical Fourier transforms in the symmetric space $D = \mathrm{Aut}(D)/K$. It follows from the Harish-Chandra formula (see §1.8, Chapter II), the compactness of the support of $S_K\chi_\Omega$, and the Paley-Wiener theorem that the Fourier transform $\widehat{S_K\chi_\Omega}$ defined on the Cartan subalgebra $\mathfrak{H} \cong \mathbb{R}^n$ of the Lie algebra of $\mathrm{Aut}(D)$ can be extended to an entire function in $\mathbb{C}^p$. Therefore, the set of zeros of $\widehat{S_K\chi_\Omega}$ in $\mathbb{R}^n$ has measure zero. Taking Fourier transforms of (3), we get $(\widehat{S_K g}) \cdot \widehat{S_K\chi_\Omega} = 0$. It now follows that $\widehat{S_K g} = 0$ almost everywhere, hence $S_K g = 0$. Since the argument remains valid for translates $g \circ \omega^{-1}$ as well, we have $g = 0$.

3.2. The result to be proved next is similar to Theorem 3.1, except that we replace the domain Ω by a compact set of arbitrary nature, and instead of integral conditions we assume the existence of approximation by complex polynomials. See §§7.8, 7.9, Chapter III, for a similar criterion in the case $n = 1$.

THEOREM. *Let D be a classical symmetric domain in $\mathbb{C}^n$, and F a compact subset of D. Suppose that the uniform closure $P(F)$ of the polynomials in $z_1, \dots, z_n$ is not $C(F)$. Let $f \in C(D)$ be a function such that $f|_{\omega(F)} \in P(\omega(F))$, $f|_{\omega(F^*)} \in P(\omega(F^*))$, where $F^* = \{\overline{z} : z \in F\}$, $\omega \in \mathrm{Aut}(D)$. Then f is holomorphic in D.*

PROOF. Let us prove first that, for every point $z_0 \in D$ there exists an element $k \in K$ such that $kz_0 = \overline{z}_0$, where K is the stationary subgroup of the origin. This can be proved directly for symmetric domains of each of the four types (see [15, §4, Chapter III]).

Let $D = D^I_{p,q}$. Express a matrix $z_0 \in D$ as $z = u_1 d v_1$, where u_1 and v_1 are unitary and d is a real matrix. Now set $k(z) = uzv^T$, where $u = \overline{u}_1 u_1^{-1}$, $v = v_1^{-1}\overline{v}_1$.

If $D = D^{II}_p$ or $D = D^{III}_p$, a matrix $z_0 \in D$ may be represented as $z_0 = usu^T$, $u \in U(p)$, s a real matrix [30]. In that case set $k(z) = (\overline{uu}^T)z(\overline{uu}^T)$.

Now consider the case $D = D^{IV}_p$, and let $z_0 \in D$. Set $\lambda = e^{i\varphi}$, where

$$\varphi = \arctan \frac{2(\mathrm{Re}\, z\,,\, \mathrm{Im}\, z)}{\|\mathrm{Im}\, z\|^2 - \|\mathrm{Re}\, z\|^2}.$$

Then the vector $w_0 = \lambda^{1/2} z_0$ satisfies the equation $(\mathrm{Re}\, w_0, \mathrm{Im}\, w_0) = 0$. There exist an orthogonal transformation $u : \mathbb{R}^p \longrightarrow \mathbb{R}^p$ such that $u(\mathrm{Re}\, w_0) = \mathrm{Re}\, w_0$, $u(\mathrm{Im}\, w_0) = -\mathrm{Im}\, w_0$, i.e., $uw_0 = \overline{w}_0$. Then the necessary element of K is given by $z \longrightarrow \lambda u z$.

Let $Y(F)$ denote the algebra of all functions that satisfy the assumption of the theorem, and $Y_K(F)$ the subalgebra of $Y(F)$ consisting of the K-invariant functions. It follows from the assumption that, for any $f \in Y_K(F)$, the function $z \longrightarrow \overline{f(\overline{z})}$ belongs to $Y_K(F)$ as well. By the above argument, K-invariance implies $f(\overline{z}) = f(z)$, and therefore $\overline{f} \in Y_K(F)$. Thus, the algebra of polynomials in the translates $f \circ \omega$, $f \in Y_K(F)$, $\omega \in \mathrm{Aut}(D)$, is closed under complex conjugation. Hence Y does not separate points of F, because otherwise, by the Stone-Weierstrass theorem $Y|_F$ would be dense in $C(F)$ and since $Y|_F \subset P(F)$ by assumption, it would follow that $P(F) = C(F)$, contradicting the definition of F.

Thus, Y is an $\mathrm{Aut}(D)$-invariant family of elements of $C(D)$ which does not separate points of D. By Lemma 4.1, Chapter I, $Y = \mathbb{C}$. Then $Y_K(F) = \mathbb{C}$.

Let f satisfy the assumption, $f \in C^2(D)$. Set

$$\widetilde{f}(z) = \int_K f(kz)\, dk.$$

Then $\widetilde{f} \in Y_K(F)$; therefore, $\widetilde{f} = \text{const}$. In particular, $f(0) = \widetilde{f}(0) = \int_K f(kz)\, dk$. Applying the Aut-invariant Laplace operator $\widetilde{\Delta}$ to both sides of this equality and setting $z = 0$, we obtain $(\widetilde{\Delta} f)(0) = 0$. Since the functions $z_k f$, $k = 1, \dots, n$, also satisfy the assumption, we have $(\widetilde{\Delta} z_k f)(0) = 0$. Hence $\partial f(0)/\partial \overline{z}_k = 0$ (see §2.1, Chapter III). Repeating the argument for the functions $f \circ \omega^{-1}$, $\omega \in \mathrm{Aut}(D)$, we see that f is holomorphic in D.

The case $f \in C(D)$ is reduced to the smooth case by the standard regularization procedure, using convolutions.

If the symmetric domain D is the unit ball $B^n \subset \mathbb{C}^n$ and F is a smooth curve bounding of a holomorphic image of the unit disk, contained in D together with a neighbourhood, we can drop the condition of approximation by complex polynomials on complex conjugate compact sets.

3.3. Proposition. *Let $D = B^n$, $\gamma = \varphi(\partial B^1)$, where $\varphi : B^1 \longrightarrow \mathbb{C}^n$ is a holomorphic map, $\varphi \in C^1(\overline{B}^1)$. Let $f \in C^1(B^n)$ be such that $f|_{\omega(\gamma)} \in P(\omega(\gamma))$ for all $\omega \in \mathrm{Aut}(B^n)$. Then f is holomorphic in B^n.*

The proof employs the same argument as in §3.2. The only difference is in the proof that a differentiable $U(n)$-invariant function satisfying the assumption of the proposition is a constant. This follows from the argument principle for analytic functions and from the fact that the quotient space $B^n/U(n)$ is one-dimensional.

For further references we formulate this statement separately.

3.4. LEMMA. *Let $f \in C^1(\gamma)$ have a holomorphic extension to $\varphi(B^1)$. If f depends only on $|z|$, it is constant on γ.*

PROOF. Denote by $\tilde{f}$ the holomorphic extension of f to $R = \varphi(B^1)$. We will show that $\tilde{f}(R) \subset f(\gamma)$. Suppose the contrary: there exists $z_0 \in R$ such that $f(z) \neq \tilde{f}(z_0)$ for all $z \in \gamma$. Then the function $g = \tilde{f} - \tilde{f}(z_0)$ has no zeros on γ. Hence, $g \circ \varphi$ does not vanish on the unit circle ∂B^1. But g depends only on $|z|$, so the index of the map $g \circ \varphi : \partial B^1 \longrightarrow \mathbb{R} \longrightarrow \mathbb{C} \setminus \{0\}$ is zero. But $g \circ \varphi$ vanishes at the point $\varphi^{-1}(z_0)$ in B^1, contrary to the argument principle.

Thus, $\tilde{f}(R) \subset f(\gamma)$. If $f \neq \text{const}$, then, since $\tilde{f}$ is an open map, $f(\gamma)$ must contain an open set, which is impossible because γ and f are smooth. Therefore, f is constant on γ.

REMARKS. Similar criteria can be formulated for functions defined in $\mathbb{C}^n$. For example, Proposition 3.3 has the following analogue for an arbitrary domain in $\mathbb{C}^n$:

Let Γ be a family of analytic submanifolds with smooth boundaries contained in D together with their closures. Assume that for every point z there exists $\gamma \in \Gamma$ such that $z \in \partial\gamma$ and $u(\gamma) \in \Gamma$ for any complex rotation u around γ. Let $f \in C^1(D)$ be such that $f|_{\partial\gamma}$ can analytically be extended to γ for any $\gamma \in \Gamma$. Then f is holomorphic in D.

If Γ consists of manifolds of complex dimension greater than one, there is no need to average over stationary transformation groups. The proof follows simply from the fact that, when $\dim_{\mathbb{C}} \gamma > 1$, the existence of holomorphic extension from the boundary of γ implies the tangent Cauchy-Riemann equations. In that case it suffices to assume that for every point $z \in D$ there exist n linearly independent complex directions, each of which is tangent to some $\gamma \in \Gamma$, that contains z.

The following holomorphy test in $\mathbb{C}$ uses a family of curves invariant with respect to translations only. Let $f \in C^1(\mathbb{R}^2)$. Suppose that, for some closed contour $\gamma \subset \mathbb{C}$ and any $a \in \mathbb{C}$, the restriction $f|_{\gamma+a}$ can be extended analytically into the interior of the contour $\gamma + a$. If $\partial f/\partial \bar{z}$ is bounded in $\mathbb{C}$, then f is an entire function. In order to prove this, one uses Green's formula to deduce convolution equations from the orthogonality of $f|_{\gamma+a}$ to the complex polynomials. One then writes that the Fourier transforms of the functions $z^k \chi_\Omega$, $k = 0, 1, 2, \ldots$, where χ_Ω is the characteristic function of the domain bounded by γ, have no common zeros in $\mathbb{C}$.

§4. Holomorphy on a unitary-invariant family of curves in a spherical layer in $\mathbb{C}^n$

In this section we prove the holomorphy test in a spherical layer (i.e., the domain between two concentric balls) in $\mathbb{C}^n$, in terms of the existence of holomorphic extension from the boundaries of analytic disks in a unitarily invariant family whose union is the spherical layer. In particular, this result is stronger than Proposition 3.3.

Let R be a holomorphic curve in $\mathbb{C}^n$, i.e., $R = \varphi(B^1)$, where B^1 is the unit disk in $\mathbb{C}$ and $\varphi : B^1 \longrightarrow \mathbb{C}^n$ a holomorphic map, $\varphi \in C^1(\overline{B}^1)$. Let γ denote the Shilov boundary $\gamma = \varphi(\partial B^1)$. Set

$$\Omega = \bigcup_{u \in U(n)} u(\gamma),$$

where $u(\gamma) = \{u(z) : z \in \gamma\}$ is the image of γ under the unitary transformation u. The set Ω is a spherical layer:

$$\Omega = \left\{ \min_{z\in\gamma} |z| \le |z| \le \max_{z\in\gamma} |z| \right\}.$$

4.1. THEOREM. *Assume that the following conditions hold:*

(i) $0 \notin R \cup \gamma$,

(ii) *γ is not contained in any complex straight line through the origin in $\mathbb{C}^n$.*

Let $f \in C^1(\Omega)$ be such that for any $u \in U(n)$

$u \in U(n)$ the restriction $f|_{u(\gamma)}$ may be extended holomorphically to $u(R)$. $(*)$

Then f is holomorphic inside Ω.

The proof uses harmonic analysis in the unit sphere S in $\mathbb{C}^n$ (see [26]). Let $H(p, q)$ denote the space of homogeneous harmonic polynomials in $\mathbb{C}^n$ of total degree p in $z_1, \dots, z_n$ and of total degree q in $\overline{z}_1, \dots, \overline{z}_n$. The restrictions of $H(p, q)$ to S are mutually orthogonal, and $L^2(S)$ decomposes into the direct sum of these subspaces. The group $U(n)$ acts by translations in $H(p, q)$. Let $K_{pq} \in H(p, q)$ denote a spherical function which is invariant with respect to the stationary subgroup $U_e \subset U(n)$ of the point $e = (1, 0, \dots, 0)$ and $\|K_{pq}\|^2_{L^2(S)} = K_{pq}(e)$. The spherical functions are characterized by the equation

$$K_{pq}(e) \int_{U_e} K_{pq}(uks)\, dk = K_{pq}(ue) K_{pq}(s), \qquad u \in U(n), \ s \in S. \tag{1}$$

The function K_{pq} defines the kernel of the orthogonal projection operator from $L^2(S)$ onto $H(p, q)$:

$$\left(\pi_{pq} f\right)(s) = \int_{U(n)} f(ue) K_{pq}(u^{-1}s)\, ds, \qquad s \in S.$$

4.2. Lemma. *If f satisfies condition $(*)$, then the functions*

$$f_{pq}(rs) = c_{pq}(s)K_{pq}(s), \qquad s \in S,$$

where

$$c_{pq}(r) = \frac{1}{K_{pq}(e)} \int\limits_{U(n)} f(rue)K_{pq}(u^{-1}e)\,du \tag{2}$$

also satisfy $(*)$.

Proof. We easily deduce from (1) that

$$f_{pq}(rs) = \int\limits_{U(n)} f(ru^{-1}s)K_{pq}(ue)\,du.$$

The function on the right obviously satisfies $(*)$.

4.3. Lemma. *The function f belongs to the closed linear span Z in $L^2(\Omega)$ of the functions $(f \circ v)_{pq} \circ u$, $u, v \in U(n)$, $p, q \geq 0$.*

Proof. Let $R(p, q)$ denote the linear span of the functions $c(r)h(s)$, where $c(r)$ is a radial function, $h(s) \in H(p, q)$, such that $c(r)h(s) \in L^2(\Omega)$. The space $L^2(\Omega)$ is the direct sum of $R(p, q)$. Consider the orthogonal projection $P_e : L^2(\Omega) \longrightarrow L_e^2(\Omega)$ onto the subspace of U_e-invariant functions:

$$P_e f = \int\limits_{U_e} (f \circ k)\,dk = \sum_{p,q\geq 0} f_{pq}.$$

Let $L_e^2(\Omega) = Z \oplus Z^{\perp}$ be an orthogonal decomposition. If $f = f_1 + f_2$, $f_1 \in Z$, $f_2 \in Z^{\perp}$, then $P_e f \in Z$, and since Z is $U(n)$-invariant, also $P_e f_1 \in Z$. Hence $P_e f_2 \in Z$ and so $f_2 \perp P_e f_2$. Since P_e is an orthogonal projection, it follows that $P_e f_2 = 0$. The same argument applied to $f \circ v$, $v \in U_{(n)}$, yields $P_e(f_2 \circ v) = 0$, i.e., $\int_{U(n)} f_2(vkz)\,dk = 0$, $v \in U(n)$, $z \in \Omega$. Hence, considering the projections of f_2 onto $R(p, q)$, we see that

$$\begin{aligned}(\pi_{pq} f_2)(rve) &= \int\limits_{U(n)} f_2(rue)K_{pq}(u^{-1}ve)\,du \\ &= \int\limits_{U(n)} \int\limits_{U_e} f_2(vkr\eta e)dk \cdot K_{pq}(\eta^{-1}e)\,d\eta = 0\end{aligned}$$

for all $p, q \geq 0$, i.e., $f_2 = 0$. Hence $f \in Z$.

4.4. Lemma. *If all the functions*

$$z_1^{k_1} \cdots z_n^{k_n} g(z), \qquad k_1 + \cdots + k_n = N,$$

satisfy condition $(*)$ *for some* $N = 0, 1, \ldots$, *then so does the function* $g(z)$.

PROOF. It follows from (i) that functions $\varphi_1^u, \ldots, \varphi_n^u$ such that

$$u(R) = \left\{z_1 = \varphi_1^u(t), \ldots, z_n = \varphi_n^u(t), t \in \overline{B}^1\right\},$$

have no common zeros in the disk $\overline{B^1}$. Therefore, there exist holomorphic functions $\beta_1^u, \ldots, \beta_n^u \in C^1(\overline{B}^1)$ such that

$$\beta_1^u \varphi_1^u + \cdots + \beta_n^u \varphi_n^u \equiv 1$$

in $\overline{B}^1$. Then

$$\sum_{k_1+\cdots+k_n=N} \left[\left(\beta_1^u \varphi_1^u\right)(t)\right]^{k_1} \cdot \cdots \cdot \left[\left(\beta_n^u \varphi_n^u\right)(t)\right]^{k_n} g(\varphi^u(t))$$

$$= \left[\left(\beta_1^u \varphi_1^u\right)(t) + \cdots + \left(\beta_n^u \varphi_n^u\right)(t)\right]^N g(\varphi^u(t)) = g(\varphi^u(t));$$

hence, g satisfies condition $(*)$.

PROOF OF THEOREM 4.1. It follows from (ii) that there exists a unitary transformation u such that the contour $u(\gamma)$ is not contained in any of the sets

$$M(r_1, \ldots, r_n) = \{z \in \mathbb{C}^n : |z_i| = r_i|z|, \ i = 1, \ldots, n\}.$$

Indeed, let $z^0 \in \gamma$. Pick $u_1 \in U(n)$ such that $z^1 = u_1(z^0)$ is not on any of the hypersurfaces $z_i = 0$, $i = 1, \ldots, n$. If the image $u_1(\gamma)$ of γ is contained in some set $M(r_1, \ldots, r_n)$, then, considering a nondiagonal unitary matrix u_2 such that $u_2(z^1) = z^1$, we see that $(u_2 u_1)(\gamma)$ is not contained in any of the above sets (e.g., the point $z^2 \in u_1(\gamma)$ which does not lie on the complex straight line through 0 and z^1 is mapped by u_2 onto a point which is not in $M(r_1, \ldots, r_n)$, $r_i = |z_i^1|/|z^1|$). Since the assumptions of the theorem are unitarily invariant, we may assume from the start that γ is not contained in any of the sets $M(r_1, \ldots, r_n)$.

Fix p, q, $q \geq 1$. By Lemma 4.3, f_{pq} satisfies condition $(*)$. Since this condition is invariant with respect to unitary transformations and the space $H(p, q)$ is the uniform closure of the $U(n)$-cyclic linear span of K_{pq}, it follows that $(*)$ holds for any function $c_{pq}(r)h(s)$, where $s \in S$, $h \in H(p, q)$, and c_{pq} is given by formula (2). Set

$$g_1(rs) = c_{pq}(r) s_1^p \overline{s}_2^q, \qquad g_2(rs) = c_{pq}(r) s_2^p \overline{s}_1^q.$$

As multiplication by powers of z_1, z_2 preserves condition $(*)$, we may assume without loss of generality that p and q are relatively prime, so there exist positive integers k and l such that $kq - lp = 1$.

We can represent $g_1^k g_2^l$ as

$$g_1^k(rs) g_2^l(rs) = c(r) F(s), \qquad c(r) = c_{pq}^{l+k}(r), \quad F(s) = |s_1|^{2lp} |s_2|^{2lq} s_2^m \overline{s}_1,$$

where $m = kp - lq > 0$. Consider the element $h(s) = s_2^m \overline{s}_1$ of $H(m, 1)$. Since

$$\int_S F(s)\overline{h(s)}\, d\sigma(s) > 0,$$

it follows that $F(s)$ has a nonzero projection onto $H(m, 1)$. Therefore, the functions

$$\alpha_1(rs) = c(r)s_2^m \overline{s}_1, \qquad \alpha_i(rs) = c(r)s_1^m \overline{s}_i, \quad i = 2, \dots, n,$$

satisfy condition $(*)$; hence, the same is true for the function

$$\begin{aligned}(rs_1)^{m+1}\alpha_1(rs) &+ \sum_{i=2}^{n}(rs_2)^m(rs_i)\alpha_i(rs)\\ &= c(r)r^{m+1}s_1^m s_2^m = c(|z|)|z|^{1-m}z_1^m z_2^m, \quad z = rs.\end{aligned}$$

We now apply the same argument to linear combinations of translations by unitary transformations. Since $H(2m, 0)$ is irreducible, condition $(*)$ holds for the functions

$$c(|z|)|z|^{1-m}z_1^{k_1}\cdots z_n^{k_n}, \qquad k_1 + \cdots + k_n = 2m.$$

It follows from Lemmas 4.4 and 3.4 that $c(|z|)$ must have the form $c(|z|) = c_0 \cdot |z|^{m-1}$. Hence, $\alpha_1(rs) = c_0 \cdot r^{m-1}s_2^m\overline{s}_1$. Together with the functions $c_0 r^{m-1} f_{i,\overline{k}}(s)$, where $\overline{k} = (k_1, \dots, k_n)$, $k_1 + \cdots + k_n = m-1$, $i = 1, \dots, n$,

$$f_{i,\overline{k}}(s) = s_1^{k_1}\cdots s_n^{k_n}\left[|s_i|^2\sum_{j\neq i}(k_j+1) - (k_i+1)\sum_{j\neq i}|s_j|^2\right]$$

also satisfies condition $(*)$, since $f_{i,\overline{k}} \in H(m, 1)$. By Lemma 4.4 this is also true for the functions

$$c_0\left(|s_i|^2\sum_{j\neq i}(k_j+1) - (k_i+1)\sum_{j\neq i}|s_j|^2\right) = c_0\left[(n+m-1)|s_i|^2 - k_i - 1\right]$$

and, therefore, also for the functions $|s_i|^2 = |z_i|^2/|z|^2$, $i = 1, \dots, n$. Being real valued, the functions $|z_i|^2/|z|^2$, $i = 1, \dots, n$, are constant on γ. Since γ is not contained in any of the sets $M(r_1, \dots, r_n)$, this is possible only if $c_0 = 0$, which means that $c_{pq} = 0$. Thus, $f_{pq} = 0$ if $q \geq 1$. Now let $q = 0$. Then $f_{p,0}(rs) = c_{p,0}(r)s_1^p$. As above, it follows from Lemmas 4.4 and 3.4 that $c_{p,0}(r) = c_0 r^p$; hence, $f_{p,0}(z) = c_0 z^p$. By Lemma 4.3, f is the $L^2(\Omega)$-limit of a sequence of holomorphic polynomials, and therefore f itself is holomorphic in Ω.

REMARK. The proof also includes the case when γ is contained in a sphere rS about the origin. In that case the spherical layer degenerates into a sphere, and the theorem asserts that f has a holomorphic extension from the sphere

rS into the complex ball bounded by the sphere. A particular case of this statement, when γ is the section of the sphere by a complex straight line not containing zero, was proved by Rudin [26].

EXAMPLES. (1) The statement is false if γ contains the origin. For example, consider the map $\varphi_\varepsilon : B^1 \longrightarrow \mathbb{C}^2$ defined by

$$\varphi_\varepsilon(t) = (t^2 + \varepsilon t\,,\, t)\,, \qquad \varepsilon > 0\,,$$

and the function $f(z) = z_1^2 \overline{z}_2$. For any unitary transformation $u \in U(2)$, the restriction of f to $u(\gamma)$ is

$$\begin{aligned}&\left[u_{11}(t^2 + \varepsilon t) + u_{12}t\right]^2 \left[\overline{u}_{21}(\overline{t}^2 + \varepsilon \overline{t}) + \overline{u}_{22}\overline{t}\right] \\ &\quad = \left[u_{11}(t + \varepsilon) + u_{12}\right]^2 \left[\overline{u}_{21}(1 + \varepsilon t) + \overline{u}_{22}t\right]\,, \qquad t\overline{t} = 1\,,\end{aligned}$$

which can obviously be extended into $u(R)$. However, f is not holomorphic in the domain

$$\Omega_\varepsilon = \{(2 - 2\varepsilon + \varepsilon^2)^{1/2} \le |z| \le (2 + 2\varepsilon + \varepsilon^2)^{1/2}\}.$$

When $\varepsilon = 0$, this yields an example of a function defined on the sphere of radius $\sqrt{2}$ about the origin, which has a holomorphic extension to the analytic manifolds $u(R)$ (containing zero) but no extension into the complex ball.

(2) The necessity of condition (ii) is even more obvious. As an example, take a real nonconstant function which is constant on the intersection of Ω with each complex straight line through the origin.

§5. The one-dimensional holomorphic extension property

In this section we discuss one-dimensional holomorphic extension, already mentioned in Chapter II, §2, Theorem 2.2.5, and Chapter III, §2, Remark to Theorem 2.2.

Let G be a bounded domain in $\mathbb{C}^n$ with a smooth boundary ∂G. Let Γ be a family of complex straight lines in $\mathbb{C}^n$. We say that G is *sufficient* for the holomorphic extension from ∂G, if any function $f \in C(\partial G)$ with the property: for any $\Lambda \in \Gamma$, the restriction $f|_{\Lambda \cap \partial G}$ has a continuous holomorphic extension to $\Lambda \cap G$, may be extended continuously to G as a holomorphic function.

It was first noted in [31] that the family of all complex straight lines that intersect the unit ball $B^n \subset \mathbb{C}^n$ is sufficient for the holomorphic extension from the complex sphere. Afterwards Stout [96] used the complex Radon transform in order to extend this result to arbitrary bounded domains with C^2-boundary.

The proofs, in both [31] and [96], essentially use only the Cauchy theorem in each complex one-dimensional "slice". This indicates that the family of straight lines is overly large. Indeed, Rudin [26] later proved that in the case of a ball the family of straight lines cutting the ball and located at a fixed distance from the center of the ball is sufficient. Even "smaller" sufficient families were found by Globevnik.(2)

Below we present some results strengthening Stout's theorems. Using an idea due to Kytmanov [16], we will apply a projective version of the Martinelli-Bochner formula to prove a theorem about one-dimensional holomorphic extension.(3)

5.1. THEOREM. *Let G be a bounded domain in $\mathbb{C}^2$, $\partial G \in C^2$. Let V be an open subset of G, and let Γ_V denote the family of complex straight lines that intersect V. Then Γ_V is sufficient for holomorphic extension from the boundary of G.*

PROOF. Let $f \in C(\partial G)$ possess the one-dimensional holomorphic extension property with respect to the family Γ_V.

Let $a \in V$. The value of the Martinelli-Bochner integral at a can be obtained by integrating over the projective space of Cauchy integrals in the sections $\Lambda \cap \partial G$ by complex straight lines Λ through a:

$$I[f](a) = \int_{\partial G} f(\zeta) \sum_{k=1}^{n} (-1)^k \frac{|\zeta_k - a_k|}{|\zeta - a|^{2n}} \, d\zeta \wedge d\bar{\zeta}\,[k]$$

$$= \int_{\mathbb{CP}^{n-1}} \int_{(a+\Lambda)\cap\partial G} f(t) \frac{dt}{t} \wedge \omega(\zeta), \tag{1}$$

where t is a complex parameter along Λ such that $t = 0$ corresponds to a and ω is the Fubini-Study differential form in $\mathbb{CP}^{n-1}$.

Let $h \in A(G)$. Since the Cauchy integral in each section $(a + \Lambda) \cap \partial G$ is multiplicative with respect to functions holomorphic in the section, it follows from (1) and from the assumptions on f that

$$I[fh](a) = h(a)I[f](a).$$

Since the Martinelli-Bochner integral is a harmonic function, it follows that $I[f]$ and $hI[f]$ are harmonic on V, and hence $I[f]$ is holomorphic on V. Then $\partial I[f]/\partial \bar{z}_k = 0$ everywhere in G, since the functions $\partial I[f]/\partial \bar{z}_k$, $k = 1, \ldots, n$, are harmonic in G and vanish in an open set V, i.e., $I[f]$ is a holomorphic function. It now follows from a theorem of Kytmanov [44] that the boundary value of $I[f]$ coincides with those of f, so it is the holomorphic extension of f.

The next proposition claims that in domains whose complement is connected, an arbitrarily small shift of a family of parallel complex straight lines is sufficient. The proof is in fact contained in Kytmanov's argument.

(2)*Remark for the English edition*: For further results in this direction see Globevnik and Stout [109].

(3)A similar idea was used already in Chapter III, Theorems 2.1 and 2.2.

5.2. Theorem. *Let G be a bounded domain whose complement is a connected set. Let U be an open subset of the unit sphere $S \subset \mathbb{C}^n$. Let Π_U denote the family of complex straight lines $\{a + ts, \ t \in \mathbb{C}\}$, $a \in \mathbb{C}^n$, $s \in U$, that intersect G. Then Π_U is sufficient for holomorphic extension from ∂G.*

Proof. Let $f \in C(\partial D)$ possess the one-dimensional holomorphic extension property for Π_U. Fix a unit vector $e \in U$. It is clear that for sufficiently small $\delta > 0$ and sufficiently large $\lambda > 0$ the ball $B = B_\delta(\lambda e)$ of radius δ about λe does not intersect G, and every complex straight line Λ that intersects both G and $B_\delta(\lambda e)$ belongs to Π_U.

Let $a \in B$. By Sard's theorem, for almost all $\Lambda \in \Pi_U$ the section $\Lambda \cap \partial G$ is a closed smooth curve. Since a is not contained in G, it follows from the Cauchy theorem that the inner integral in (1) (with respect to t) vanishes for almost all Λ. Therefore, $I[f](a) = 0$ for $a \in B$. Harmonicity implies $I[f](z) = 0$ for $z \in \mathbb{C}^n \setminus \overline{G}$. Hence [16, Theorem 2.4.4] $f \in A(\partial G)$.

If one retains the full family of complex sections, one can weaken the assumptions on f: instead of the existence of holomorphic extension to sections it will suffice to assume that the integrals over the curves in the boundary of G vanish.

5.3. Proposition. *Let $\omega = \sum_{k=1}^n \omega_k \, dz_k$ be a $(1, 0)$-form holomorphic in a neighborhood of the unit sphere $S \subset \mathbb{C}^n$, $\omega \neq 0$ on S. Let $f \in C^1(S)$ be such that for some r, $0 < r < 1$, and for any complex straight line Λ intersecting the unit ball B^n but disjoint from the ball rB^n*

$$\int_{\Lambda \cap S} f\omega = 0. \tag{1}$$

Then $f \in A(S)$.

Proof. Pick an arbitrary point $z_0 \in S$. By applying a unitary transformation, we may assume that $z_0 = (0, 0, \dots, 0, i)$. We use standard parameters in a neighborhood of z_0 in S: if $z_k = x_{2k-1} + ix_{2k}$, $k = 1, \dots, n$, then the local coordinates are $t_i = x_i$, $i = 1, \dots, 2n-1$. Subtracting $f(z_0)$ from $f(z)$, we may assume that $f(z_0) = 0$. Condition (1) still holds, because by the Stokes theorem $\int_{\Lambda \cap B^n} \omega = 0$, since ω is closed. Set

$$A(t) = \begin{pmatrix} \partial f(t)/\partial t_1 & & & \dots & & & \partial f/\partial t_{2n-1} \\ 1 & i & 0 & \dots & & & 0 \\ & & & \dots & & & \\ 0 & \dots & & 1 & i & 0 & 0 \\ 0 & & & \dots & & 0 & 1 \end{pmatrix}.$$

By assumption,

$$\int_0^{2\pi} e^{i\varphi} (f\omega_k)(z_1, \dots, z_{k-1}, e^{i\varphi} z_k, z_{k+1}, \dots, z_n) \, d\varphi = 0,$$

$$k = 1, \dots, n-1, \ |z_n| > r,$$

or, using the coordinates $t_1, \dots, t_{2n-1}$,

$$\int_0^{2\pi} e^{i\varphi}(f\omega_k)(t_1, \dots, t_{2k-1}\cos\varphi - t_{2k}\sin\varphi, t_{2k-1}\sin\varphi + t_{2k}\cos\varphi, \dots, t_{2n-1})\,d\varphi = 0.$$

Differentiating this equality with respect to t_{2k-1}, setting $t = 0$, and taking into account that $f(z_0) = 0$, we obtain

$$\int_0^{2\pi} e^{i\varphi}\left[\frac{\partial(f\omega_k)}{\partial t_{2k-1}}(0)\cos\varphi + \frac{\partial(f\omega_k)}{\partial t_{2k}}(0)\sin\varphi\right]d\varphi = \pi\omega_k(z_0)\left[\frac{\partial f}{\partial t_{2k-1}}(0) + i\frac{\partial f}{\partial t_{2k}}(0)\right] = 0, \qquad k = 1, \dots, n.$$

Since $\omega_k(z_0) \neq 0$, we have $\partial f(0)/\partial t_{2k-1} + i\partial f(0)/\partial t_{2k} = 0$, and this means that the rank of $A(0)$ is zero, i.e., $df \wedge dz_1 \wedge \dots \wedge dz_n = 0$ at z_0. Since z_0 is an arbitrary point, the Bochner-Severi theorem yields $f \in A(S)$.

REMARK. If the form ω is constant on S, the proposition holds for $f \in C(S)$, since in that case condition (1) is invariant with respect to unitary changes of variables, and one can apply regularization procedures to f, using convolutions with smooth functions in $U(n)$ functions.

REMARKS TO CHAPTER IV. Propositions 1.2 and 1.6 were proved by Rudin [26], for the special case of a complex ball.

Theorem 2.1 answers a question of Zalcman [106] about the Morera theorem for conformally invariant families of contours. The results of §2 have a bearing on the Pompeiu problem (for details see [91], [92]).

Theorem 2.7 was also obtained independently by Berenstein and Zalcman [57]. Theorem 3.1 was modified by Berenstein [56], who replaced conditions on $\partial f/\partial \bar{z}_k$ by the assumption that $\partial\Omega$ is nonanalytic instead. Berenstein's result is valid only for a complex ball. His proof relies on reducing the Pompeiu problem to an inverse problem for an overdetermined Neumann system [40], [42], [43].

In the particular case $D = B^n$, Theorem 3.2 was proved by Rudin [90] and independently by Gichev [40] under weaker assumptions (not assuming the existence of approximation on the conjugate family of compact sets) as a corollary of the classification of Möbius algebras in an open complex ball.

Theorem 4.1 was proved jointly by the author and Semenov. In the particular case $n = 1$, when Ω is a flat annular domain, the theorem was proved by Globevnik [71], using the ideas from [31].

The one-dimensional holomorphic extension property was studied in detail by Globevnik. In the case of the ball $B^2 \subset \mathbb{C}^2$ he found fairly small families of complex straight lines that are sufficient for analytic extension from the sphere $S^2 \subset \mathbb{C}^2$. Grinberg [73] studied the relation between the one-dimensional holomorphic extension property and the complex Radon transform used by Stout in [96].

Theorem 5.2 strengthens results of Globevnik [69] describing the holomorphic extension from a sphere in $\mathbb{C}^2$.

The one-dimensional holomorphic extension theorem holds also for balls in Banach spaces (Druzkowski [63]).

References

1. M. L. Agranovskiĭ, *Invariant algebras on the boundaries of symmetric domains*, Dokl. Akad. Nauk SSSR **197** (1971), no. 1, 9–11; English transl., Soviet Math. Doklady **12** (1971), 371–374.
2. ———, *Invariant algebras on noncompact symmetric Riemannian spaces*, Dokl. Akad. Nauk SSSR **207** (1972), no. 3, 513–516; English transl. in Soviet Math. Doklady **13** (1972).
3. ———, *On antisymmetric algebras of differentiable functions on smooth manifolds*, Sibirsk. Mat. Zh. **18** (1977), no. 2, 455–458; English transl., Siberian Math. J. **18** (1977), 327–331.
4. ———, *Fourier transform on $SL_2(\mathbb{R})$ and Morera type theorems*, Dokl. Akad. Nauk SSSR **243** (1978), no. 6, 1353–1356; English transl., Soviet Math. Doklady **19** (1978), 1522–1525.
5. ———, *Invariant function algebras on homogeneous spaces of nocompact semisimple Lie groups*, Mat. Zametki **28** (1980), no. 5, 645–652; English transl., Math. Notes **28** (1980), 779–783.
6. ———, *Tests for holomorphy in symmetric domains*, Sibirsk. Mat. Zh. **22** (1981), no. 2, 7–18; English transl., Siberian Math. J. **22** (1981), 171–179.
7. ———, *Function spaces in a disk invariant under multiplication by z and conformal translations*, Sibirsk. Mat. Zh. **22** (1981), no. 3, 3–8; English transl. in Siberian Math. J. **22** (1981).
8. ———, *Invariant spaces and traces of holomorphic functions on Shilov boundaries of the classical domains*, Sibirsk. Mat. Zh. **25** (1984), no. 2, 3–12; English transl. in Siberian Math. J. **25** (1984).
9. ———, *Invariant function algebras in symmetric spaces*, Trudy Moskov. Mat. Obshch. **47** (1984), 158–178; English transl. in Trans. Moscow Math. Soc. **47** (1985), 175–197.
10. ———, *Invariant function spaces and algebras on Lie groups and symmetric domains*, Colloquia Math. Soc. Janoš Bolyai (A. Haar Memorial Conf., Budapest, 1985), vol. 49, Akademiai Kiado, Budapest, 1987, pp. 115–126.
11. ———, *Affine-invariant function algebras on the Heisenberg group*, Dokl. Akad. Nauk SSSR **289** (1986), no. 2, 265–268; English transl., Soviet Math. Doklady **34** (1987), 53–56.
12. ———, *Invariant function spaces on the Heisenberg group*, Sibirsk. Mat. Zh. **28** (1987), no. 3, 6–27; English transl., Siberian Math. J. **28** (1987), 358–375.
13. ———, *Poisson integral in Möbius-invariant spaces in the complex ball*, Siberian Math. J. **31** (1990), no. 6; English transl. in Siberian Math. J. **31** (1990).
14. M. L. Agranovskiĭ and A. M. Semenov, *Analyticity on unitarily invariant families of curves in $\mathbb{C}^n$*, Sibirsk. Mat. Zh. **29** (1988), no. 1, 192–196; English transl., Siberian Math. J. **29** (1988), 149–152.
15. ———, *$SU(m, n)$-invariant spaces of holomorphic functions*, Preprint TR90-2, University of Maryland, 1990, pp. 1–24.
16. L. A. Aĭzenberg and A. P. Yuzhakov, *Integral representations and residues in multidimensional complex analysis*, "Nauka", Novosibirsk, 1988; English transl. of the 1st edition, Amer. Math. Soc., Providence, RI, 1983.

17. A. Erdélyi (ed.), *Higher transcendental functions*, McGraw-Hill, New York, 1954.
18. V. S. Vladimirov, *Methods of the theory of function of several complex variables*, "Nauka", Moscow, 1984; English transl. of the 1st edition, MIT Press, Cambridge, MA, 1966.
19. T. W. Gamelin, *Uniform algebras*, Prentice-Hall, Englewood Cliffs, NJ, 1969.
20. D. P. Zhelobenko, *Compact Lie groups and their representations*, "Nauka", Moscow, 1970; English transl., Amer. Math. Soc., Providence, RI, 1973.
21. D. P. Zhelobenko and A. I. Shtern, *Representations of Lie groups*, "Nauka", Moscow, 1983. (Russian)
22. A. A. Kirillov, *Elements of representation theory*, "Nauka", Moscow, 1972; English transl., Springer-Verlag, Berlin and New York, 1976.
23. M. M. Lavrent'ev, V. G. Romanov, and S. P. Shishatskiĭ, *Ill posed problems of mathematical physics and analysis*, "Nauka", Moscow, 1980; English transl., Amer. Math. Soc., Providence, RI, 1986.
24. A. Barut and R. Ronczka, *Theory of group representations and applications*, vol. 2, PWN, Warszawa, 1977.
25. I. I. Pyatetskiĭ-Shapiro, *Geometry of the classical domains and theory of automorphic functions*, "Nauka", Moscow, 1961; English transl., Gordon and Breach, New York, 1969.
26. W. Rudin, *Function theory in the unit ball of* $\mathbb{C}^n$, Springer-Verlag, New York, 1980.
27. G. Szegö, *Orthogonal polynomials*, 3rd edition, Amer. Math. Soc., Providence, RI, 1967.
28. S. Helgason, *Differential geometry and symmetric spaces*, Academic Press, New York, 1962.
29. E. M. Stein and G. Weiss, *Introduction to Fourier analysis on Euclidean spaces*, Princeton Univ. Press, Princeton, NJ, 1971.
30. Hua Lo-ken, *Harmonic analysis of functions of several complex variables in the classical domains*, Princeton Univ. Press, Princeton, NJ, 1963.
31. M. L. Agranovskiĭ and R. E. Val'skiĭ, *Maximality of invariant algebras of functions*, Sibirsk. Mat. Zh. **12** (1971), no. 1, 3–12; English transl., Siberian Math. J. **12** (1971), 1–7.
32. L. A. Aĭzenberg and A. Sh. Dautov, *Holomorphic functions of several complex variables with nonnegative real part. Traces of holomorphic and pluriharmonic functions on the Shilov boundary*, Mat. Sb. **99** (1976), no. 3, 342–355; English transl., Math. USSR-Sb. **28** (1976), 301–313.
33. E. B. Vinberg, S. G. Gindikin, and I. I. Pytetskiĭ-Shapiro, *Classification and canonical realization of complex bounded homogeneous domains*, Trudy Moskov. Mat. Obshch. **12** (1963); English transl. in Trans. Moscow Math. Soc. **12** (1963) 404–437.
34. V. S. Vladimirov, *Holomorphic functions with positive imaginary part in the future tube.* I, Mat. Sb. **93** (1974), 9–17; English transl., Math. USSR-Sb. **22** (1974), 1–16.
35. V. S. Vladimirov and A. G. Sergeev, *Complex analysis in the future tube*, Sovremennye Problemy Matematiki. Fundamental'nye Napravleniya, vol. 8, VINITI, Moscow, 1985, pp. 191–266; English transl. in Encyclopaedia Math. Sci., vol. 8, Springer, Berlin and New York, 1990.
36. I. M. Gel′fand, *On subrings of the ring of continuous functions*, Uspekhi Mat. Nauk **12** (1957), no. 1, 247–251; English transl., Amer. Math. Soc. Transl. Ser. 2 **16** (1960), pp. 477–479.
37. S. G. Gindikin, *Analysis in homogeneous domains*, Uspekhi Mat. Nauk **19** (1964), no. 4, 3–92; English transl. in Russian Math. Surveys **19** (1964).
38. V. M. Gichev, *Invariant function algebras on Lie groups*, Sibirsk. Mat.Zh. **20** (1979), no. 1, 23–36; English transl., Siberian Math. J. **20** (1979), 15–25.
39. ———, *Maximal ideal spaces of invariant algebras*, Funktsional. Anal. i Prilozhen. **13** (1979), no. 3, 73–76; English transl. in Functional Anal. Appl. **13** (1979).
40. ———, *Invariant algebras of continuous functions on balls and Euclidean spaces*, Sibirsk. Mat. Zh. **25** (1984), no. 4, 32–36; English transl., Siberian Math. J. **25** (1984), 534–537.
41. E. A. Gorin, *Commutative Banach algebras generated by a group of unitary elements*, Funktsional. Anal. i Prilozhen **1** (1967), no. 3, 86–87; English transl., Functional Anal. Appl. **1** (1967), 243–244.
42. E. A. Gorin and V. M. Zolotarevskiĭ, *Maximal subalgebras in algebras with involution*, Mat. Sb. **85** (1971), no. 3, 373–387; English transl., Math. USSR-Sb. **14** (1971), 367–382.

43. E. A. Gorin, *On the research of G. E. Shilov in the theory of commutative Banach algebras and its subsequent development*, Uspekhi Mat. Nauk **33** (1978), no. 4, 169–188; English transl., Russian Math. Surveys **33** (1978), 193–217.
44. A. M. Kytmanov, *Holomorphy criterion for integrals of Martinelli-Bochner type*, Combinatorial and Asymptotic Analysis, "Nauka", Krasnoyarsk, 1975, pp. 169–177. (Russian)
45. N. K. Nikolskiĭ, *Invariant subspaces in operator theory and function theory*, Itogi Nauki i Techniki, Matematicheskiĭ Analiz, vol. 12, VINITI, Moscow, 1974, pp. 199–412; English transl. in J. Soviet Math. **5** (1976), no. 1/2.
46. P. K. Rashevskiĭ, *Description of closed invariant subspaces of some function spaces*, Trudy Moskov. Mat. Obshch. **39** (1979), 139–185; English transl. in Proc. Moscow Math. Soc. **1981**, no. 1.
47. G. M. Henkin, *Method of integral representations in complex analysis*, Sovremennye Problemy Matematiki. Fundamental'nye Napravlenuya, vol. 7, VINITI AN SSSR, Moscow, 1985, pp. 23-124; English transl., Encyclopaedia Math. Sci., vol. 7, Springer-Verlag, Berlin and New York, 1989, pp. 19–116.
48. R. O. Wells Jr., *Function theory on differentiable manifolds in* $\mathbb{C}^n$, Contributions to Analysis, Academic Press, New York, 1974, pp. 407–441.
49. E. M. Chirka, *Approximation by holomorphic functions on smooth manifolds in* $\mathbb{C}^n$, Mat. Sb. **78** (1969), no. 1, 101–123; English transl. in Math. USSR-Sb. **7** (1969).
50. G. E. Shilov, *Homogeneous rings of functions*, Uspekhi Mat. Nauk **6** (1951), no. 1, 89–135. (Russian)
51. ———, *Homogeneous rings of functions on the torus*, Dokl. Akad. Nauk SSSR **85** (1952), no. 5. (Russian)
52. J. Arazy and S. Fisher, *Some aspects of the minimal Möbius invariant space of analytic functions on the unit disk*, Interpolation spaces and related topics in analysis (Lund, 1983), Lecture Notes in Math., vol. 1070, Springer-Verlag, Berlin, 1984, pp. 24–44.
53. J. Arazy, S. Fisher, and J. Peetre, *Möbius-invariant function spaces*, J. Reine Angew. Math. **363** (1985), 110–144.
54. R. Arens, *The maximal ideals of certain function algebras*, Pacific J. Math. **8** (1958), 641–648.
55. C. A. Berenstein, *An inverse spectral theorem and its relation to the Pompeiu problem*, J. Analyse Math. **37** (1980), 128–144.
56. ———, *A test for holomorphy in the unit ball of* $\mathbb{C}^n$, Proc. Amer. Math. Soc. **90** (1984), no. 1, 88–90.
57. C. A. Berenstein and L. Zalcman, *Pompeiu's problem on symmetric cpaces*, Comment. Math. Helv. **55** (1980), 593–621.
58. C. A. Berenstein and M. Shahshahani, *Harmonic analysis and the Pompeiu problem*, Amer. J. Math. **105** (1983), no. 5, 1217–1229.
59. A. G. Brandstein, *Compact 2-manifolds as maximal ideal spaces*, Proc. Amer. Math. Soc. **41** (1973), no. 2, 438–450.
60. J.-E. Björk, *Compact groups operating on Banach algebra*, Math. Ann. **205** (1973), no. 4, 281–297.
61. J. Delsart, *Note sur une propriété nouvelle des fonctions harmoniques*, C. R. Acad. Sci. Paris Ser. I Math. **246** (1958), 1358–1360.
62. J. Delsart and J. L. Lions, *Moyennes generalisées*, Comment. Math. Helv. **33** (1959), 59–69.
63. L. M. Druzkowski, *Continuous holomorphic extension from the boundary in Banach space*, Proc. Intern. Conf. on Complex Anal. and Appl., Varna, 1981, pp. 157–160.
64. M. Educhi, M. Hashirime, and K. Okamoto, *The Paley-Wiener theorem for distributions on symmetric space*, Hiroshima Math. J. **3** (1973), no. 1, 109–120.
65. M. Freeman, *Uniform approximation on a real analutic manifold*, Trans. Amer. Math. Soc. **143** (1969), 545–553.
66. G. B. Folland and E. M. Stein, *Hardy spaces on homogeneous groups* M, Math. Notes **28**, Princeton Univ. Press, Princeton, NJ, 1982.
67. R. Gangolli, *Invariant function algebras on compact semisimple Lie groups*, Bull. Amer. Math. Soc. **71** (1965), no. 3, 634–637.

68. J. Globevnik, *On boundary values of holomorphic functions on balls*, Proc. Amer. Math. Soc. **85** (1982), 61–64.
69. ———, *On holomorphic extensions from spheres in* $\mathbb{C}^2$, Proc. Roy. Soc. Edinburg **94A** (1983), 113–120.
70. ———, *A family of lines for testing holomorphy in the ball of* $\mathbb{C}^2$, preprint, vol. 24, Ljubljana Univ., Ljubljana, 1986.
71. ———, *Analyticity on rotation invariant families of curves*, Trans. Amer. Math. Soc. **280** (1983), 247–254.
72. D. Geller, *Fourier analysis on the Heisenberg group*, Proc. Nat. Acad. Sci. U.S.A. **74** (1977), 1328–1331.
73. E. Grinberg, *Boundary values of holomorphic functions, Radon transform and the one dimensional extension property*, preprint, Temple University, 1985.
74. ———, *A boundary analogue of Morera's theorem in the unit ball of* $\mathbb{C}^n$, Proc. Amer. Math. Soc. **102** (1988), no. 1, 114–116.
75. L. R. Hunt and R. O. Wells, Jr., *Extension of CR-functions*, Amer. J. Math. **98** (1976), 805–820.
76. L. Hörmander and J. Wermer, *Uniform approximation on compact sets in* $\mathbb{C}^n$, Math. Scand. **23** (1968), no. 1, 5–21.
77. K. D. Johnson, *On a ring of invariant polynomials on a Hermitian symmetric space*, J. Algebra **67** (1980), 72–81.
78. S. Kobayashi, *Fixed points of isometries*, Nagoya Math. J. **13** (1958), 63–68.
79. A. Koranyi, *Poisson integral and boundary components of symmetric spaces*, Invent. Math. **34** (1976), 19–34.
80. K. de Leeuw and H. Mirkil, *Intrinsic algebras on the torus*, Trans. Amer. Math. Soc. **81** (1956), 320–330.
81. ———, *Translation-invariant funcion algebras on abelian groups*, Bull. Soc. Math. France **88** (1960), 345–370.
82. ———, *Rotation-invariant algebras on the* n*-sphere*, Duke Math. J. **30** (1961), no. 4, 667–672.
83. ———, *Algebras of differentiable functions in the plane*, Ann. Inst. Fourier (Grenoble) **13** (1963), no. 2, 75–90.
84. A. Nagel and W. Rudin, *Möbius-invariant function spaces on balls and spheres*, Duke Math. J. **43** (1976), 841–865.
85. R. D. Ogden and S. Vagi, *Harmonic analysis on a nilpotent group and function theory of Siegel domains of type.* II, Adv. in Math. **33** (1979), no. 1, 31–92.
86. J. Peetre, *Invariant function spaces—a rapid survey*, technical report (1985), Lund.
87. ———, *Paracommutators and minimal spaces*, Operators and function theory, Reidel, Dordrecht, 1985, pp. 163–224.
88. D. Rider, *Translation-invariant Dirichlet algebras on compact groups*, Proc. Amer. Math. Soc. **17** (1966), no. 5, 977–985.
89. W. Rudin, *Unitarily invariant algebras of continuous functions on spheres*, Houston J. Math. **5** (1979), 253–265.
90. ———, *Möbius-invariant algebras in balls*, Ann. Inst. Fourier (Grenoble) **33** (1983), no. 2, 19–41.
91. ———, *Eigenspaces of invariant Laplacian in* B^n, J. Analyse Math. **43** (1983/1984), 136–148.
92. L. A. Rubel, *Möbius-invariant spaces of continuous fictions*, Bull. Greek Math. Soc. **20** (1979), 94–97.
93. ———, *Harmonic analysis of harmonic functions in the plane*, Proc. Amer. Math. Soc. **54** (1976), no. 1, 146–148.
94. L. A. Rubel and A. L. Shields, *Invariant subspaces of* L^∞ *and* H^∞, J. Reine Angew. Math. **272** (1975), 32–44.
95. L. A. Rubel and R. M. Timoney, *An extremal property of the Bloch space*, Proc. Amer. Math. Soc. **75** (1979), 45–49.
96. E. L. Stout, *The boundary values of holomorphic functions of several complex variables*, Duke Math. J. **44** (1977), 105–108.

97. W. Schmid, *Die Randwerte holomorpher Funktionen und hermitische symmetrischen Räumen*, Invent. Math. **9** (1969/70), 61–80.
98. L. Schwartz, *Théorie générale des fonctions moyenne-périodiques*, Ann. of Math. **42** (1947), 857–929.
99. M. Takeuchi, *Polynomial representations associated with bounded symmetric domains*, Osaka J. Math. **10** (1973), 441–445.
100. R. M. Timoney, *Maximal invariant spaces of analytic functions*, Indiana Univ. J. Math. **31** (1982), 651–664.
101. J. A. Tirao, *Self-adjoint function spaces on Riemannian symmetric spaces*, Proc. Amer. Math. Soc. **24** (1970), 223–228.
102. J. Wermer, *Approximation on a disk*, Math. Ann. **155** (1964), 331–333.
103. ———, *Polynomially convex disks*, Math. Ann. **158** (1965), 6–10.
104. J. A. Wolf, *Selfadjoint function spaces on Riemann symmetric spaces*, Proc. Amer. Math. Soc. **113** (1964), 299–315.
105. ———, *Translation-invariant function algebras on compact groups*, Pacific J. Math. **15** (1965), no. 3, pp. 1093–1099.
106. L. Zalcman, *Analyticity and the Pompeiu problem*, Arch. Rational Mech. Anal. **47** (1972), 237–254.
107. ———, *Offbeat integral geometry*, Amer. Math. Monthly **87** (1980), no. 3, 161–175.

References added for the English edition

108. Y. Beniamini and Y. Weit, *Harmonic analysis of spherical functions on $SU(1,1)$*, Ann. Inst. Fourier (Grenoble) **42** (1992), no. 3, 671–694.
109. Y. Globevnik and E. L. Stout, *Boundary Morera theorems for holomorphic functions of several complex variables*, Duke Math. J. **64** (1991), 571–615.
110. M. L. Agranovskiĭ and A. M. Semenov, *Boundary analogues of Hartogs' theorem*, Siberian Math. J. **32** (1991), no. 1.
111. C. Berenstein, D. C. Chang, D. Pascus, and L. Zalcman, *Variation on the theorem of Morera*, Contemporary Math. (1992, to appear).
112. M. L. Agranovskiĭ, C. Berenstein, D. C. Chang, and D. Pascus, *A Morera type theorem for L^2-functions in the Heisenberg group*, J. Analyse Math. **57** (1991).
113. ———, *Théorémes de Morera et Pompeiu pour le groupe de Heisenberg*, C. R. Acad. Sci. Paris Sér. I Math. (1992, to appear).
114. M. L. Agranovskiĭ, C. Berenstein, and D. C. Chang, *Morera theorem for holomorphic H^p-spaces in the Heisenberg group*, J. Reine Angew. Math. (to appear).
115. S. Thangavelu, *Spherical means and CR functions on the Heisenberg group*, J. Analyse Math. (to appear).
116. J. Faraut and A. Koranyi, *Function spaces and reproducing kernels on bounded symmetric domain*, J. Funct. Anal. **88** (1990), no. 1, 64–89.
117. J. Arazy, *Realization of the invariant inner product on the highest quotients of the composition series*, Ark. Mat. **30** (1992), no. 1, 1–24.

Recent Titles in This Series

(Continued from the front of this publication)

91 **Mamoru Mimura and Hirosi Toda,** Topology of Lie groups, I and II, 1991
90 **S. L. Sobolev,** Some applications of functional analysis in mathematical physics, third edition, 1991
89 **Valeriĭ V. Kozlov and Dmitriĭ V. Treshchëv,** Billiards: A genetic introduction to the dynamics of systems with impacts, 1991
88 **A. G. Khovanskiĭ,** Fewnomials, 1991
87 **Aleksandr Robertovich Kemer,** Ideals of identities of associative algebras, 1991
86 **V. M. Kadets and M. I. Kadets,** Rearrangements of series in Banach spaces, 1991
85 **Mikio Ise and Masaru Takeuchi,** Lie groups I, II, 1991
84 **Đáo Trọng Thi and A. T. Fomenko,** Minimal surfaces, stratified multivarifolds, and the Plateau problem, 1991
83 **N. I. Portenko,** Generalized diffusion processes, 1990
82 **Yasutaka Sibuya,** Linear differential equations in the complex domain: Problems of analytic continuation, 1990
81 **I. M. Gelfand and S. G. Gindikin, Editors,** Mathematical problems of tomography, 1990
80 **Junjiro Noguchi and Takushiro Ochiai,** Geometric function theory in several complex variables, 1990
79 **N. I. Akhiezer,** Elements of the theory of elliptic functions, 1990
78 **A. V. Skorokhod,** Asymptotic methods of the theory of stochastic differential equations, 1989
77 **V. M. Filippov,** Variational principles for nonpotential operators, 1989
76 **Phillip A. Griffiths,** Introduction to algebraic curves, 1989
75 **B. S. Kashin and A. A. Saakyan,** Orthogonal series, 1989
74 **V. I. Yudovich,** The linearization method in hydrodynamical stability theory, 1989
73 **Yu. G. Reshetnyak,** Space mappings with bounded distortion, 1989
72 **A. V. Pogorelev,** Bendings of surfaces and stability of shells, 1988
71 **A. S. Markus,** Introduction to the spectral theory of polynomial operator pencils, 1988
70 **N. I. Akhiezer,** Lectures on integral transforms, 1988
69 **V. N. Saliĭ,** Lattices with unique complements, 1988
68 **A. G. Postnikov,** Introduction to analytic number theory, 1988
67 **A. G. Dragalin,** Mathematical intuitionism: Introduction to proof theory, 1988
66 **Ye Yan-Qian,** Theory of limit cycles, 1986
65 **V. M. Zolotarev,** One-dimensional stable distributions, 1986
64 **M. M. Lavrent′ev, V. G. Romanov, and S. P. Shishat·skiĭ,** Ill-posed problems of mathematical physics and analysis, 1986
63 **Yu. M. Berezanskiĭ,** Selfadjoint operators in spaces of functions of infinitely many variables, 1986
62 **S. L. Krushkal′, B. N. Apanasov, and N. A. Gusevskiĭ,** Kleinian groups and uniformization in examples and problems, 1986
61 **B. V. Shabat,** Distribution of values of holomorphic mappings, 1985
60 **B. A. Kushner,** Lectures on constructive mathematical analysis, 1984
59 **G. P. Egorychev,** Integral representation and the computation of combinatorial sums, 1984
58 **L. A. Aĭzenberg and A. P. Yuzhakov,** Integral representations and residues in multidimensional complex analysis, 1983

(See the AMS catalog for earlier titles)